The Binomial Theorem

A Self-Study Guide to Mathematics

Volume 1

First Edition

Jianlun Xu

Copyright© 2017 by Jianlun Xu

All rights reserved. No part of this book may be reproduced, stored in a retrieval system or transmitted in any form or by any means without the prior written permission of the author.

Printed by CreateSpace, An Amazon.com Company

ISBN: 1545489645
ISBN-13: 978-1545489642

Preface

This book is written for high school students, college students, and anyone interested in teaching themselves math. The goal of the book is to help you establish a solid foundation of mathematics for your advanced studies and the preparation of SAT and ACT.

The book is one of a set of books, "*A Self-Study Guide to Mathematics*". Each book covers one particular subject of high school or college mathematics. The characteristics of this book include

- Extensive coverage of a particular math subject
- Plain language to explain math concept
- Plenty of math proofs and explanations to tell you why
- Abundant detailed examples to tell you how-to step by step

Someone may say "I know what I learned from my math class but I don't know why in detail and how to solve new math problems step by step." If this sounds like you this book is for you.

Some students feel mathematics boring because they may not be trained to think in a mathematical way during their studies. When you complete this book, you will be gradually trained to think logically through plenty of proofs and examples in detail. You will find that math likes a fun game. That reminds me of a dialogue between my daughter and I. When she was in middle school she played piano and also took part in math contests. One day she asked me "Do you think that someone good at piano will also be good at math?" "Why do you ask such question?" I wondered. "In my class a boy is good at math and competes against me in a math contest. He plays piano very well too". "That is quite possible, math is harmonious like music". I answered. Yes, math is beautiful.

The set of books, "*A Self-Study Guide to Mathematics*", is your math tutor at home. Good luck in your math study and the preparation of SAT and ACT.

I appreciate the support from my wife and daughter who make this book possible.

Jianlun Xu

Contents

1 Principle of the Binomial Theorem

1.1 Binomials and the Pascal's Triangle   1
  The Pascal's Triangle
  Properties of the Pascal's Triangle

1.2 Binomial Expansion   4
  Binomial Expansion of Binomials
  Binomial Coefficients of Binomial Expansion
  Properties of Binomial Expansion
  The General Term of Binomial Expansion

2 Applications of the Binomial Theorem

2.1 Finding a Particular Term of a Binomial Expansion   14
  The Term with Index Known
  The Term $x^s y^t$
  Constant Term
  Rational Coefficients and Terms
  The Term with the Maximum Binomial Coefficient
  The Term with the Maximum Coefficient

2.2 Finding the Sum of a Sequence   38

2.3 Proving Identities of Combinations   43

2.4 Proving Inequalities   47

2.5 Factoring Algebraic Expressions   50

2.6 Finding General Term of a Polynomial Expansion   55

2.7 Approximate Computation   58
  Simplified Formula of Approximate Computation

1 Principle of the Binomial Theorem

∎ 1.1 Binomials and Pascal's Triangle

▶ The Pascal's Triangle

A binomial is a polynomial that has two terms like $(x+y)^n$ ($n \in N$). Let's look at a few binomials and their expansions as below.

$$(x+y)^0 = 1$$
$$(x+y)^1 = x+y$$
$$(x+y)^2 = x^2 + 2xy + y^2$$
$$(x+y)^3 = x^3 + 3x^2y + 3xy^2 + y^3$$
$$(x+y)^4 = x^4 + 4x^3y + 6x^2y^2 + 4xy^3 + y^4$$
$$(x+y)^5 = x^5 + 5x^4y + 10x^3y^2 + 10x^2y^3 + 5xy^4 + y^5$$
$$(x+y)^6 = x^6 + 6x^5y + 15x^4y^2 + 20x^3y^3 + 15x^2y^4 + 6xy^5 + y^6$$
$$\cdots \qquad \cdots$$

From the above we know that the expansion of a binomial will become more complicated and tedious as its power n increases. It becomes very difficult to expand a binomial with high power in this way. Thus we need to find a new method to expand a binomial with high power. After rearranging the expansions above we have the following triangle pattern of the expansion of binomials.

$$
\begin{aligned}
(x+y)^0 &= 1 \\
(x+y)^1 &= x+y \\
(x+y)^2 &= x^2 + 2xy + y^2 \\
(x+y)^3 &= x^3 + 3x^2y + 3xy^2 + y^3 \\
(x+y)^4 &= x^4 + 4x^3y + 6x^2y^2 + 4xy^3 + y^4 \\
(x+y)^5 &= x^5 + 5x^4y + 10x^3y^2 + 10x^2y^3 + 5xy^4 + y^5 \\
(x+y)^6 &= x^6 + 6x^5y + 15x^4y^2 + 20x^3y^3 + 15x^2y^4 + 6xy^5 + y^6 \\
\cdots & \qquad \cdots
\end{aligned}
$$

Look at the pattern of highlighted coefficients above we can get a triangle of these coefficients, called the **Pascal's triangle**, as below.

$(x+y)^0$: 1
$(x+y)^1$: 1 1
$(x+y)^2$: 1 2 1
$(x+y)^3$: 1 3 3 1
$(x+y)^4$: 1 4 6 4 1
$(x+y)^5$: 1 5 10 10 5 1
$(x+y)^6$: 1 6 15 20 15 6 1
...

From the Pascal's triangle, we find that every coefficient is equivalent to a corresponding number of combinations. For example,

- $(x+y)^2 = x^2 + 2xy + y^2$

  Because $_2C_0 = {_2C_2} = 1$ and $_2C_1 = 2$, the third row of the Pascal's triangle can be replaced by these numbers of combinations.

  $$(x+y)^2 = x^2 + 2xy + y^2$$
  $$\ 1 \quad\ \ 2 \quad\ 1$$
  $$\ _2C_0 \quad _2C_1 \quad _2C_2$$
  $$(x+y)^2 = {_2C_0}x^2 + {_2C_1}xy + {_2C_2}y^2.$$

- $(x+y)^3 = x^3 + 3x^2y + 3xy^2 + y^3$

  Because $_3C_0 = {_3C_3} = 1$ and $_3C_1 = {_3C_2} = 3$, the fourth row of the Pascal's triangle can be replaced by these numbers of combinations.

  $$(x+y)^3 = x^3 + 3x^2y + 3xy^2 + y^3$$
  $$\ 1 \quad\ \ 3 \quad\ \ 3 \quad\ 1$$
  $$\ _3C_0 \quad _3C_1 \quad _3C_2 \quad _3C_3$$
  $$(x+y)^3 = {_3C_0}x^3 + {_3C_1}x^2y + {_3C_2}xy^2 + {_3C_3}y^3.$$

We can continue to write the expansion of a binomial of power n in this way. Then the coefficients of expansion of binomials, called **binomial coefficients**, can be expressed by the number of combinations. The Pascal's triangle can be rewritten by the number of combinations.

$(x+y)^0$: 1
$(x+y)^1$: $_1C_0$ $_1C_1$
$(x+y)^2$: $_2C_0$ $_2C_1$ $_2C_2$
$(x+y)^3$: $_3C_0$ $_3C_1$ $_3C_2$ $_3C_3$
$(x+y)^4$: $_4C_0$ $_4C_1$ $_4C_2$ $_4C_3$ $_4C_4$
$(x+y)^5$: $_5C_0$ $_5C_1$ $_5C_2$ $_5C_3$ $_5C_4$ $_5C_5$
$(x+y)^6$: $_6C_0$ $_6C_1$ $_6C_2$ $_6C_3$ $_6C_4$ $_6C_5$ $_6C_6$
...

$+$ ◀ *symmetrical axis*

▶ Properties of the Pascal's Triangle

1) The terms of the nth row are the binomial coefficients of the binomial expansion of a binomial $(x+y)^n$.
2) There are $n + 1$ terms in the nth row.
3) The first term $_nC_0$ and the last term $_nC_n$ of every row are equal to 1.

$$_nC_0 = _nC_n = 1$$

4) Two terms of a row are equal if they are symmetrical about the symmetrical axis of the Pascal's triangle. In other words, they have the same distance to their the nearest end of the row.

$$_nC_m = _nC_{n-m} \quad (m = 1, 2, 3, \ldots, n-1)$$

5) Each inner term in a row is the addition of two terms right above it.

$$_{n+1}C_m = _nC_{m-1} + _nC_m$$

6) The middle term(s) of each row is the maximum binomial coefficient in that row.
   - When n is even number there is a single middle term. For example, the fifth row ($n = 4$)

     1 4 **6** 4 1

   - When n is odd number there are two middle terms. For example, the sixth row ($n = 5$)

     1 5 **10 10** 5 1

Example 1.1.1   The Pascal's Triangle

Given two rows of the Pascal's triangle as below,

(Row n) $_nC_0$ $_nC_1$ $_nC_2$ $_nC_3$ $_nC_4$ $_nC_5$ a $_nC_7$
(Row $n+1$) $_{n+1}C_0$ b $_{n+1}C_2$ c $_{n+1}C_6$ d

find
1) n, a, and the binomial represented by the first row
2) b, c, and d
3) the maximum term(s) of each row

Solution:
1) The first row has 8 terms then $n = 7$ and $a = {_7}C_6$. The row represents the binomial $(x+y)^8$.
2) By the property 4 of the Pascal's triangle, $b = {_7}C_0 + {_7}C_1$ and $c = {_7}C_3 + {_7}C_4$. And by the property 3 of the Pascal's triangle, $d = {_8}C_8 = 1$.
3) In the row n ($n = 7$), the term $_nC_3$ and the term $_nC_4$ are two middle terms that are the maximum terms in the row. In the row $n + 1$ ($n + 1 = 8$), the term $c = {_7}C_3 + {_7}C_4$ is the middle term and the maximum term in the row.

1.2 Binomial Expansion

▶ Binomial Expansion of Binomials

Generally we can express the binomial expansions of a binomial of power n by binomial coefficients as below.

$$(x+y)^0 = 1$$
$$(x+y)^1 = {}_1C_0\, x + {}_1C_1\, y$$
$$(x+y)^2 = {}_2C_0\, x^2 + {}_2C_1\, xy + {}_2C_2\, y^2$$
$$(x+y)^3 = {}_3C_0\, x^3 + {}_3C_1\, x^2 y + {}_3C_2\, xy^2 + {}_3C_3\, y^3$$
$$(x+y)^4 = {}_4C_0\, x^4 + {}_4C_1\, x^3 y + {}_4C_2\, x^2 y^2 + {}_4C_3\, xy^3 + {}_4C_4\, y^4$$
$$\cdots$$

$$\underbrace{(x+y)^n}_{\text{a binomial of power } n} = \underbrace{{}_nC_0\, x^n + {}_nC_1\, x^{n-1} y + \cdots + {}_nC_k\, x^{n-k} y^k + \cdots + {}_nC_{n-1}\, xy^{n-1} + {}_nC_n\, y^n}_{\text{binomial expansion}}$$

We can obtain above binomial expansion by expanding a binomial of power n using the Pascal's triangle. The left side of the formula above is a binomial of power n and the right side is called the **binomial expansion** of the binomial.

The method to expand a binomial using the number of combinations as its binomial coefficients is called the **binomial theorem**. The binomial theorem makes it easier to expand a binomial of higher power.

The Binomial Theorem

A binomial, $(x+y)^n$, can be expanded as

$$(x+y)^n = {}_nC_0\, x^n + {}_nC_1\, x^{n-1} y + \cdots + {}_nC_k\, x^{n-k} y^k + \cdots + {}_nC_n\, y^n$$

$$= \sum_{k=0}^{n} {}_nC_k\, x^{n-k} y^k \qquad (n, k \in N^+, 0 \leq k \leq n). \qquad (1.2.1)$$

The coefficient of the general term of the binomial expansion, the $(k+1)^{\text{th}}$ term, is

$${}_nC_k = \frac{n!}{(n-k)!\,k!}.$$

1.2 Binomial Expansion

Proof   The Binomial Theorem

Prove the binomial $(x+y)^n = \sum_{k=0}^{n} {}_nC_k\, x^{n-k} y^k$ $(n, k \in N^+, 0 \leq k \leq n)$.

Proof:
We use mathematical induction to prove the binomial theorem.

1) Let $P(n)$ be the equation above.

$$P(n): (x+y)^n = \sum_{k=0}^{n} {}_nC_k\, x^{n-k} y^k$$

2) Induction:
- When $n = 1$,

$$P(1): (x+y)^1 = x + y = {}_1C_0\, x + {}_1C_1\, y$$

thus $P(1)$ is true.

- When $n = m$,

$$P(m): (x+y)^m = \sum_{k=0}^{m} {}_mC_k\, x^{m-k} y^k.$$

Assume $P(m)$ is true then we have the following equation when $n = m + 1$.

$P(m+1)$: $(x+y)^{m+1} = (x+y)^m (x+y)$

$$= \left(\sum_{k=0}^{m} {}_mC_k\, x^{m-k} y^k \right)(x+y)$$

The $(k + 1)^{th}$ term, the general term of $P(m + 1)$ becomes two parts

$${}_mC_k\, x^{m-k} y^k \cdot (x+y) = {}_mC_k\, x^{m-k+1} y^k + {}_mC_k\, x^{m-k} y^{k+1} \qquad (a)$$

Similarly we have the kth term and the $(k + 2)^{th}$ term of $P(m + 1)$.

$${}_mC_{k-1}\, x^{m-k+1} y^{k-1} \cdot (x+y) = {}_mC_{k-1}\, x^{m-k+2} y^{k-1} + {}_mC_{k-1}\, x^{m-k+1} y^k \qquad (b)$$

$${}_mC_{k+1}\, x^{m-k-1} y^{k+1} \cdot (x+y) = {}_mC_{k+1}\, x^{m-k} y^{k+1} + {}_mC_{k+1}\, x^{m-k-1} y^{k+1} \qquad (c)$$

By the property of combinations we have

$$\begin{cases} {}_mC_{k-1} + {}_mC_k = {}_{m+1}C_k \\ {}_mC_k + {}_mC_{k+1} = {}_{m+1}C_{k+1} \end{cases}$$

Then the first part of the $(k + 1)^{th}$ term (a) will be combined with the second part of the kth term (b) to form the $(k +1)^{th}$ term of $(x + y)^{m+1}$. Similarly the second part of the $(k + 1)^{th}$ term (a) and the first part of the $(k + 2)^{th}$ term (c) are combined to form the $(k + 2)^{th}$ term of $(x + y)^{m+1}$. In this way we find all rest terms of $(x + y)^{m+1}$. The figure below is straight forward.

the kth term of $(x+y)^m$ the $(k+1)^{th}$ term of $(x+y)^{km}$ the $(k+2)^{th}$ term of $(x+y)^m$

$$\cdots + {}_mC_{k-1}x^{m-k+1}y^{k-1}\cdot(x+y) + {}_mC_k x^{m-k}y^k\cdot(x+y) + {}_mC_{k+1}x^{m-k-1}y^{k+1}\cdot(x+y) + \cdots$$

$$\cdots + {}_mC_{k-1}x^{m-k+2}y^k + {}_mC_{k-1}x^{m-k+1}y^{k} + {}_mC_k x^{m-k+1}y^k + {}_mC_k x^{m-k}y^{k+1} + {}_mC_{k+1}x^{m-k}y^{k+1} + \cdots$$

$$\cdots + ({}_mC_{k-1} + {}_mC_k)x^{m-k+1}y^k \quad + \quad ({}_mC_k + {}_mC_{k+1})x^{m-k}y^{k+1} + \cdots$$

$$\cdots + {}_{m+1}C_k x^{m+1-k}y^k \quad + \quad {}_{m+1}C_{k+1}x^{m-k}y^{k+1} + \cdots$$

the $(k+1)^{th}$ term of $(x+y)^{m+1}$ the $(k+2)^{th}$ term of $(x+y)^{m+1}$

Then

$$(x+y)^{m+1} = {}_{m+1}C_0 x^{m+1} + {}_{m+1}C_1 x^m y + \cdots + {}_{m+1}C_k x^{m+1-k}y^k + \cdots + {}_{m+1}C_{m+1}y^{m+1}.$$

$$(x+y)^{m+1} = \sum_{k=0}^{m+1} {}_{m+1}C_k x^{m+1-k}y^k$$

Therefor $P(m+1)$ is true.

3) Conclusion:

$P(n)$ is true $\forall n \in N^+$, i.e. $(x+y)^n = \sum_{k=0}^{n} {}_nC_k x^{n-k}y^k$ is true $\forall n \in N^+$.

▶ Binomial Coefficients of Binomial Expansions

From the binomial theorem we know that a binomial $(x+y)^n$ can be expanded as

$$(x+y)^n = {}_nC_0 x^n + {}_nC_1 x^{n-1}y + \cdots + {}_nC_k x^{n-k}y^k + \cdots + {}_nC_n y^n.$$

Where $n, k \in N^+$ and $0 \leqslant k \leqslant n$.

The numbers of combinations associated with the terms in the binomial expansion are called **binomial coefficients**. They are arranged in the order as

$${}_nC_0, {}_nC_1, \cdots, {}_nC_k, \cdots, {}_nC_{n-1}, {}_nC_n.$$

The followings are the properties of binomial coefficients of a binomial expansion.

1. The binomial coefficient of a term in a binomial expansion is the number

1.2 Binomial Expansion

of combinations corresponding to that term.

2. The sum of all binomial coefficients is 2^n.

$$_nC_0 + {_nC_1} + \cdots + {_nC_k} + \cdots + {_nC_{n-1}} + {_nC_n} = 2^n.$$

3. The sum of binomial coefficients of all terms with odd index is equal to the sum of binomial coefficients of all terms with even index.

$$_nC_1 + {_nC_3} + {_nC_5} + \cdots = {_nC_0} + {_nC_2} + {_nC_4} + \cdots = 2^{n-1}.$$

4. By the property of combinations, $_nC_m = {_nC_{n-m}}$, we have

$$_nC_0 = {_nC_n},\ {_nC_1} = {_nC_{n-1}},\ {_nC_2} = {_nC_{n-2}},\ \ldots,\ {_nC_k} = {_nC_{n-k}}.$$

Thus two binomial coefficients are equal if they have the same distance to their nearest end of the binomial expansion. In other words, they are symmetrical about the symmetrical axis of the binomial expansion.

5. Binomial coefficient increases from the beginning of a binomial expansion until the middle of the binomial expansion then decreases to the end. The maximum binomial coefficient is associated with the middle term(s) of the binomial expansion.

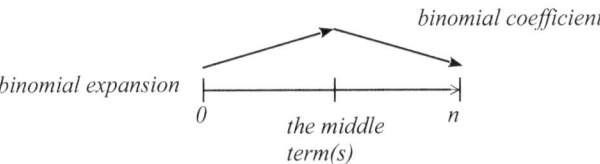

6. The maximum binomial coefficient(s) in a binomial expansion is located at the middle of the binomial expansion.

- If the power n of a binomial is an even number, the middle term of the binomial expansion is the $(\frac{n}{2}+1)^{th}$ term. Its binomial coefficient is

$$_nC_{\frac{n}{2}}$$

and it is the maximum binomial coefficient in a binomial expansion.

- If the power n of the binomial is a odd number, both the $(\frac{n+1}{2})^{th}$ term and the $(\frac{n+1}{2}+1)^{th}$ term are the middle terms of the binomial expansion. Their binomial coefficient are the same and both are the maximum binomial coefficients.

$$_nC_{\frac{n-1}{2}} = {_nC_{\frac{n+1}{2}}}.$$

▶ Properties of Binomial Expansions

For a binomial of power n like

$$(x+y)^n = {}_nC_0 x^n + {}_nC_1 x^{n-1} y + \cdots + {}_nC_k x^{n-k} y^k + \cdots + {}_nC_n y^n,$$

we have the properties of binomial expansion of the binomial above as below.

1. A binomial expansion is an identity.
2. The left side of the formula (a) is a binomial of power n and the right side is the binomial expansion of the binomial.
3. There are total $n + 1$ terms in the binomial expansion.
4. From the left end to the right end of the binomial expansion, the power of the term x decreases by one from n to 0 and the power of the term y increases by one from 0 to n.
5. The addition of powers of the term x and the term y of each term equals n.
6. In a binomial expansion the **coefficient** and the **binomial coefficient** of a term have different meaning.
   - The binomial coefficient of a term is a number of combinations like ${}_nC_k$. It is determined by the power of a binomial and the position of the term in the binomial expansion before combining like terms.
   - The coefficient of a term is refereed to the coefficient of the term after combining like terms.

   We give two examples to explain it. The following abbreviations are used to indicate the status of a binomial expansion.
   BEBC - the binomial expansion before combining like terms
   BEAC - the binomial expansion after combining like terms

1) $(x+5)^2 = {}_2C_0 x^2 + {}_2C_1 x \cdot (5) + {}_2C_2 \cdot (5)^2$ (BEBC)
   $= 1 \cdot x^2 + 10 \cdot x + 25$ (BEAC)

   binomial coefficients: ${}_2C_0, {}_2C_1, {}_2C_2,$
   coefficients: $1, 10, 25$
   the binomial coefficient of the term x: ${}_2C_1 = 2.$
   the coefficient of the term x: ${}_2C_1 \cdot (5) = 10.$

2) $(3x+2y)^3 = {}_3C_0 (3x)^3 + {}_3C_1 \cdot (3x)^2 \cdot (2y) + {}_3C_2 (3x)(2y)^2 + {}_3C_3 (2y)^3$ (BEBC)
   $= 27 x^3 + 54 x^2 y + 36 x y^2 + 8 y^3$ (BEAC)

   binomial coefficients: ${}_3C_0, {}_3C_1, {}_3C_2, {}_3C_3,$
   coefficients: $27, 54, 36, 8$
   the binomial coefficient of the term xy^2: ${}_3C_2 = 3.$
   the coefficient of the term xy^2: ${}_3C_2 \cdot (3) \cdot (2)^2 = 36.$

1.2 Binomial Expansion

Example 1.2.1   Binomial Expansion

Write the binomial expansion, binomial coefficient, and the coefficients of the binomial expansion for the following binomials.

1) $(x+2)^3$ 2) $(3x+4y)^4$ 3) $(5x-3y)^4$

Solution:

1) $(x+2)^3 = {}_3C_0 x^3 + {}_3C_1 x^2(2) + {}_3C_2 x(2^2) + {}_3C_3(2^3)$
$= 1x^3 + 6x^2 + 12x + 8.$
   binomial coefficients: ${}_3C_0, {}_3C_1, {}_3C_2, {}_3C_3$
           coefficients: 1, 6, 12, 8

2) $(3x+4y)^4$
$= {}_4C_0(3x)^4 + {}_4C_1(3x)^3(4y) + {}_4C_2(3x)^2(4y)^2 + {}_4C_3(3x)(4y)^3 + {}_4C_4(4y)^4$
$= 81x^4 + 432x^3 y + 72x^2 y^2 + 48x y^3 + 256 y^4.$
   binomial coefficients: ${}_4C_0, {}_4C_1, {}_4C_2, {}_4C_3, {}_4C_4$
           coefficients: 81, 432, 72, 48, 256

3) $(5x-3y)^4 = (5x+(-3y))^4$
$= {}_4C_0(5x)^4 + {}_4C_1(5x)^3(-3y) + {}_4C_2(5x)^2(-3y)^2 + {}_4C_3(5x)(-3y)^3 + {}_4C_4(-3y)^4$
$= 625x^4 - 1{,}500 x^3 y^2 + 1{,}350 x^2 y^2 - 60 x y^3 + 81 y^4.$
   binomial coefficients: ${}_4C_0, {}_4C_1, {}_4C_2, {}_4C_3, {}_4C_4.$
           coefficients: 625, –1,500, 1,350, –60, 81

Example 1.2.2   Binomial Expansion

Find the binomial expansion of the following binomials.

1) $(ax+by)^5$ 2) $(3x-2y)^4$ 3) $(\frac{x^2}{2}+\frac{1}{x})^5$

Solution:

1) $(ax+by)^5 =$
${}_5C_0(ax)^5 + {}_5C_1(ax)^4(by) + {}_5C_2(ax)^3(by)^2 + {}_5C_3(ax)^2(by)^3 + {}_5C_4(ax)(by)^4 + {}_5C_5(by)^5$
$= 1 \cdot (ax)^5 + 5 \cdot (ax)^4(by) + 10 \cdot (ax)^3(by)^2 + 10 \cdot (ax)^2(by)^3 + 5 \cdot (ax)(by)^4 + 1 \cdot (by)^5$
$= a^5 x^5 + 5a^4 b x^4 y + 10 a^3 b^2 x^3 y^2 + 10 a^2 b^3 x^2 y^3 + 5 a b^4 x y^4 + b^5 y^5.$

2) $(3x-2y)^4$
$= {}_4C_0(3x)^4 + {}_4C_1(3x)^3(-2y) + {}_4C_2(3x)^2(-2y)^2 + {}_4C_3(3x)(-2y)^3 + {}_4C_4(-2y)^4$
$= 81 \cdot x^4 - 216 \cdot x^3 y + 216 \cdot x^2 y^2 - 96 \cdot x y^3 + 16 \cdot y^4.$

3) $(\frac{x^2}{2}+\frac{1}{x})^5$

$= {}_5C_0(\frac{x^2}{2})^5 + {}_5C_1(\frac{x^2}{2})^4(\frac{1}{x}) + {}_5C_2(\frac{x^2}{2})^3(\frac{1}{x})^2 + {}_5C_3(\frac{x^2}{2})^2(\frac{1}{x})^3 + {}_5C_4(\frac{x^2}{2})(\frac{1}{x})^4 + {}_5C_5(\frac{1}{x})^5$

$= \frac{x^{10}}{2} + \frac{5}{2} \cdot x^7 + 5 \cdot x^4 + 5 \cdot x + \frac{5}{2} \cdot \frac{1}{x^2} + \frac{1}{x^5}.$

Example 1.2.3 Binomial Expansion

Find out the first four terms of the binomial expansion of the following expressions

1) $(ax+b\sqrt{y})^{20}$ 　　2) $(1+x+x^2)^4$ 　　3) $(\sqrt{x}-\dfrac{1}{\sqrt{x}})^8$

Solution:

1) $(ax+b\sqrt{y})^{20}$

$= {}_{20}C_0(ax)^5 + {}_{20}C_1(ax)^{19}(b\sqrt{y}) + {}_{20}C_2(ax)^{18}(b\sqrt{y})^2 + {}_{20}C_3(ax)^{17}(b\sqrt{y})^3 + \cdots$

$= a^5 \cdot x^5 + 20\,a^{19}bx^{19}\sqrt{y} + 190\,a^{18}b^2x^{18}y + 1{,}140\,a^{17}b^3x^{17}\sqrt{y}^3 + \cdots$.

2) $(1+x+x^2)^4$

$= ((1+x)+x^2)^4$

$= {}_4C_0(1+x)^4 + {}_4C_1(1+x)^3(x^2) + {}_4C_2(1+x)^2(x^2)^2 + {}_4C_3(1+x)(x^2)^3 + {}_4C_4(x^2)^4$

$= (1+x)^4 + 4(1+x)^3x^2 + 6(1+x)^2x^4 + 4(1+x)x^6 + x^8$

$= 1 + 4x + 10x^2 + 16x^3 + 19x^4 + 16x^5 + 10x^6 + 4x^7 + x^8$.

3) $(\sqrt{x}-\dfrac{1}{\sqrt{x}})^8$

$= {}_8C_0(\sqrt{x})^8 + {}_8C_1(\sqrt{x})^7(\dfrac{-1}{\sqrt{x}}) + {}_8C_2(\sqrt{x})^6(\dfrac{-1}{\sqrt{x}})^2 + {}_8C_3(\sqrt{x})^5(\dfrac{-1}{\sqrt{x}})^3 + \cdots$

$= {}_8C_0x^4 - {}_8C_1x^3 + {}_8C_2x^2 + {}_8C_3x + \cdots$.

Example 1.2.4 Binomial Expansion

Write out the binomial expansion of the following expressions

1) $(x^2+x+2)^4$ 　　2) $(ax^2-y)^5$

Solution:

1) $(x^2+x+2)^4 = (x^2+(x+2))^4$

$= {}_4C_0(x^2)^4 + {}_4C_1(x^2)^3(x+2) + {}_4C_2(x^2)^2(x+2)^2 + {}_4C_3(x^2)(x+2)^3 + {}_4C_4(x+2)^4$

$= x^8 + 4x^6(x+2) + 6x^4(x+2)^2 + 4x^2(x+2)^3 + (x+2)^4$ 　　(a)

Since $(x+2)^2 = x^2 + 4x + 4$

$(x+2)^3 = {}_3C_0x^3 + {}_3C_1x^2 \cdot 2 + {}_3C_2x \cdot 2^2 + {}_3C_3 \cdot 2^3$,

$(x+2)^4 = {}_4C_0x^4 + {}_4C_1x^3 \cdot 2 + {}_4C_2x^2 \cdot 2^2 + {}_4C_3x \cdot 2^3 + {}_4C_4 \cdot 2^4$

the formula (a) becomes

$x^8 + 4x^7 + 8x^6 + 4x^5 + 25x^4 + 56x^3 + 56x^2 + 32x + 16$.

2) $(ax^2-y)^5$

$= {}_5C_0(ax^2)^5 - {}_5C_1(ax^2)^4 y + {}_5C_2(ax^2)^3 y^2 - {}_5C_3(ax^2)^2 y^3 + {}_5C_4(ax^2)y^4 - {}_5C_5 y^5$

$= a^5x^{10} - (5a^4)x^8 y + (10a^3)x^6 y^2 - (10a^2)x^4 y^3 + (5a)x^2 y^4 - y^5$

$= a^5x^{10} - 5a^4x^8 y + 10a^3x^6 y^2 - 10a^2x^4 y^3 + 5ax^2 y^4 - y^5$

1.2 Binomial Expansion

Example 1.2.5 *Binomial Expansion*

Find the expansion of the following expressions
1) $(3x+1)^4+(3x-1)^4$ 2) $(1+2x)^2(1-x)^3$

Solution:

1) $(3x+1)^4+(3x-1)^4$
$= ({}_4C_0(3x)^4 + {}_4C_1(3x)^3 + {}_4C_2(3x)^2 + {}_4C_3(3x) + {}_4C_4)$
$+ ({}_4C_0(3x)^4 - {}_4C_1(3x)^3 + {}_4C_2(3x)^2 - {}_4C_3(3x) + {}_4C_4)$
$= 2({}_4C_0(3x)^4 + {}_4C_2(3x)^2 + {}_4C_4)$
$= 162x^4 + 108x^2 + 2.$

2) $(1+2x)^2(1-x)^3$
$= ({}_2C_0 + 2{}_2C_1 x + 4{}_2C_2 x^2)({}_3C_0 - {}_3C_1 x + {}_3C_2 x^2 - {}_3C_3 x^3)$
$= (1+4x+4x^2)(1-3x+3x^2-x^3)$
$= 1+x-5x^2-x^3+12x^4-4x^5.$

Example 1.2.6 *Binomial Expansion*

In the expansion of the polynomial $(x+y+z)^3$,
1) how many terms in total in the expansion after combining like terms?
2) how many terms with y^2?
3) how many terms with yz^2?
4) what is the coefficient of the term xyz?

Solution:

Arrange the polynomial $(x+y+z)^3$ to the binomial $(x+(y+z))^3$. Expand it as below.

$(x+(y+z))^3 = {}_3C_0 x^3 + {}_3C_1 x^2(y+z) + {}_3C_2 x(y+z)^2 + {}_3C_3 (y+z)^3$
$= {}_3C_0 x^3 + {}_3C_1 x^2 y + {}_3C_1 x^2 z + {}_3C_2 x y^2 + 2 \cdot {}_3C_2 x y z + {}_3C_2 x z^2$
$+ {}_3C_3 (y^3 + 3y^2 z + 3yz^2 + z^3).$

Because ${}_3C_0 = {}_3C_3 = 1$ and ${}_3C_1 = {}_3C_2 = 3$, the expansions of the polynomial $(x+y+z)^3$ becomes

$(x+y+z)^3 = x^3+3x^2 y+3x^2 z+3xy^2+6xyz+3xz^2+y^3+3y^2 z+3yz^2+z^3.$

Then we have the following answers.
1) There are ten terms in the binomial expansion after combining like terms.
2) There are two terms with y^2, $3xy^2$ and $3zy^2$.
3) There is one terms with yz^2, $3yz^2$.
4) The coefficient of the term xyz is 6.

▶ The General Term of Binomial Expansions

The general term of a binomial expansion is the $(k+1)^{th}$ term of the binomial expansion, which can be used to determine other terms in the binomial expansion.

> **DEFINITION**   **The General Term of Binomial Expansion**
>
> The general term of the binomial expansion of a binomial $(x+y)^n$ is the $(k+1)^{th}$ term, denoted by
>
> $$T_{k+1} = {_n}C_k \, x^{n-k} \, y^k \quad (n, k \in N^+, 0 \leqslant k \leqslant n) \quad\quad (1.2.2)$$
>
> The number of combinations, ${_n}C_k$, is called the **binomial coefficient** of the term T_{k+1}.

The following comments are about the general term.

- The general term of a binomial expansion is the $(k+1)^{th}$ term instead of the kth term. For instance, the 3rd term is expressed by the $(2+1)^{th}$ term and $k = 2$.
- The binomial coefficient ${_n}C_k$ of the general term T_{k+1} is determined by the power n of the binomial and the position, index k, of the term in the binomial expansion.
- If the positions of the variable x and y in the formula $(x+y)^n$ are switched it results in a different binomial expansion. In other words, the $(k+1)^{th}$ term of the binomial expansion of $(x+y)^n$ and that of $(y+x)^n$ are different.

Example 1.2.7   *General Term of Binomial Expansion*

Given a binomial $(ax+by)^{10}$ $(a, b \in R; n \in N)$, find the following terms and highlight their binomial coefficient and coefficient.
(*BC*–binomial coefficient, *TC*–the coefficient)

1) the fourth term from the left end of the binomial expansion
2) the fourth term from the right end of the binomial expansion
3) the middle term of the binomial expansion.

Solution:

Use the general term $T_{k+1} = {_n}C_k \, x^{n-k} \, y^k$ to determine other terms. Because $n = 10$ there are total eleven terms in the binomial expansion.

1.2 Binomial Expansion

1) The fourth term from the left end of the binomial expansion is
$$T_4 = T_{3+1} = \underbrace{{}_{10}C_3(ax)^{10-3}(by)^3}_{BC} = \underbrace{{}_{10}C_3 a^7 b^3 x^7 y^3}_{TC}.$$

2) The fourth term from the right end of the binomial expansion is the eighth term from the beginning of the binomial expansion
$$T_8 = T_{7+1} = \underbrace{{}_{10}C_7(ax)^{10-7}(by)^7}_{BC} = \underbrace{{}_{10}C_7 a^3 b^7 x^3 y^7}_{TC}.$$

3) The middle term of the binomial expansion is the sixth term from the beginning of the binomial expansion
$$T_6 = T_{5+1} = \underbrace{{}_{10}C_5(ax)^{10-5}(by)^5}_{BC} = \underbrace{{}_{10}C_5 a^5 b^5 x^5 y^5}_{TC}.$$

2 Applications of the Binomial Theorem

Applications of the binomial theorem cover the following aspects.

- Find a particular term in a binomial expansion
- Find the sum of a sequence
- Prove identities of combinations
- Prove inequality
- Factoring Algebraic Expressions
- Polynomial Expressions
- Approximate computation

■ 2.1 Finding a Particular Term in a Binomial Expansion

By the general term of a binomial expansion,
$$T_{k+1} = {}_nC_k x^{n-k} y^k \quad (n, k \in N^+, 0 \leqslant k \leqslant n),$$
we can find a particular term in the binomial expansion without expanding the binomial.

Based on the conditions given, we set certain values to the formula of the general term T_{k+1}. The following cases are used often.

1) The term with index known
2) The term $x^s y^t$ (s and t are known)
3) Constant term
4) Rational term
5) The term with the maximum binomial coefficient
6) The term with the maximum coefficient

We discuss these six cases and give examples to help readers to understand easily.

2.1 Finding a Particular Term in a Binomial Expansion

▶ The Term with Index Known

If the index of a term is given we can find the term by the formula of the general term T_{k+1}. Because T_{k+1} is the kth term of the expansion so we just simply set

$$k + 1 = \text{the index given.}$$

Example 2.1.1 *The Term with Index Known*

Find the specific term from the following binomial expansions.
1) The 4th term of the binomial expansion of $(2x+7y)^5$
2) The 8th term of the binomial expansion of $(5x-y)^9$
3) The 7th term of the binomial expansion of $(x+3)^{14}$

Solution:

1) Let $k+1 = 4$ and $n = 5$. $T_4 = T_{3+1} = {}_5C_3 \cdot (2x)^{5-3} \cdot (7y)^3 = 13720 \cdot x^2 \cdot y^3$.
   The binomial coefficient ${}_5C_3$
   The coefficient ${}_5C_3 \cdot 2^2 \cdot 7^3 = 13720$.

2) Let $k+1 = 8$ and $n = 9$. $T_8 = T_{7+1} = {}_9C_7 \cdot (5x)^{9-7} \cdot (-y)^7 = -900 \cdot x^3 \cdot y^7$.
   The binomial coefficient ${}_9C_7$
   The coefficient ${}_9C_7 \cdot 5^2 \cdot (-1)^7 = -900$.

3) Let $k+1 = 7$ and $n = 14$. $T_7 = T_{6+1} = {}_{14}C_6 \cdot x^{14-6} \cdot (3)^6 = 2,189,187 \cdot x^8$.
   The binomial coefficient ${}_{14}C_6 = 3,003$
   The coefficient ${}_{14}C_6 \cdot 3^6 = 2,189,187$.

Example 2.1.2 *The Term with Index Known*

1) If the coefficient of the 9th term equals that of the 11th term of the binomial expansion of the binomial $(x+1)^n$ find $n = ?$
2) The sum of the binomial coefficient of last three terms of the binomial expansion of the binomial $(1+2x)^n$ is 16, find the 3rd term of the binomial expansion.

Solution:

1) By the $(k+1)^{th}$ term, we have the 9th term and the 11th term as below.
$$T_9 = T_{8+1} = {}_nC_8 x^{n-8}$$
$$T_{11} = T_{10+1} = {}_nC_{10} x^{n-10}$$

Let ${}_nC_8 = {}_nC_{10}$ and we have
$$(n-8)(n-9) = 10 \cdot 9.$$
$$n^2 - 17n - 18 = 0.$$

Two solutions are obtained, 18 and -1. We take positive integer $n = 18$.

2) Let $\quad {}_nC_{n-2}+{}_nC_{n-1}+{}_nC_n=\dfrac{n(n-1)}{2}+n+1=16$

$$n^2+n-30=0.$$

We obtain two solutions -6 and 5. Take $n = 5$ and the general term becomes

$$T_{k+1}=2^k\cdot{}_5C_k\cdot x^k.$$

Then the 3rd term is $\quad T_3=2^2\cdot{}_5C_2\cdot x^2=40\cdot x^2.$

▶ The Term $x^s y^t$

To find a term containing $x^s y^t$ (s, t are positive integers known and $s + t = n$) in a binomial expansion, just set the power of x and y to s and t ($n - k = s$ and $k = t$) respectively in the formula of the general term T_{t+1}. As $k = t$, the general term T_{k+1} becomes

$$T_{k+1}={}_nC_t\,x^{n-t}\,y^t \qquad (n,t\in N^+, 0\leqslant t\leqslant n)$$

Example 2.1.3   The Term $x^s y^t$

Find the binomial coefficient and the coefficient of the term x^9 in the binomial expansion of the binomial $(2x-3)^{12}$.

Solution:

The general term of the binomial expansion is $T_{k+1}={}_{12}C_k\,x^{12-k}(-3)^k$. Set $12-k=9$ and $k = 3$. Then the fourth term is the term x^9.

$$T_{3+1}={}_{12}C_3\,x^{12-3}(-3)^3 = (-27\cdot{}_{12}C_3)x^9.$$

Its binomial coefficient is ${}_{12}C_3$ and coefficient $-27\cdot{}_{12}C_3$.

Example 2.1.4   The Term $x^s y^t$

Prove that $4\cdot{}_{15}C_{11}$ is the coefficient of the term of $x^{11}y^2$ in the binomial expansion of the binomial $(x-\sqrt{2y})^{15}$.

Solution:

The general term of the binomial expansion is

$$T_{k+1}={}_{15}C_k\,x^{15-k}(-\sqrt{2y})^k = {}_{15}C_k\,x^{15-k}(-1)^k\cdot 2^{\frac{k}{2}}\,y^{\frac{k}{2}}.$$

Let $\dfrac{k}{2}=2$ and $k = 4$. Then the fifth term is the term containing $x^{11}y^2$.

$$T_{4+1}={}_{15}C_4\,x^{11}(-1)^4\cdot 2^2\,y^2 = (4\cdot{}_{15}C_4)\cdot x^{11}y^2.$$

Its binomial coefficient is ${}_{15}C_4$ and coefficient $4\cdot{}_{15}C_4=4\cdot{}_{15}C_{11}$.

2.1 Finding a Particular Term in a Binomial Expansion

Example 2.1.5   The Term $x^s y^t$

If the sum of all coefficients of the binomial expansion of the binomial $(3x+1)^n$ is 256, find the coefficient of the term of x^2 in the binomial expansion.

Solution:

$$(3x+1)^n = {}_nC_0(3x)^n + {}_nC_1(3x)^{n-1} + \cdots + {}_nC_{n-1}3x + {}_nC_n$$

Let $x = 1$, then $4^n = 256$ and $n = 4$. The general term becomes

$$T_{k+1} = {}_4C_k(3x)^{4-k}.$$

Let $4 - k = 2$. and $k = 2$. We have the term x^2,

$$T_{2+1} = {}_4C_2(3x)^2 = 54 \cdot x^2.$$

Thus the coefficient of the term of x^2 is 54.

Example 2.1.6   The Term $x^s y^t$

Find the coefficient of the term x^5 in the expansion of the expression

$$(1-x^3)(1+x)^{10}.$$

Solution:

We discuss the expression $(1-x^3)(1+x)^{10}$ in two separate parts.

- Let T_{p+1} be the general term of the binomial expansion of $(1-x^3)$.

$$T_{p+1} = {}_1C_p(-1)^p(x^3)^p \qquad (p = 0, 1)$$

- Let T_{q+1} be the general term of the binomial expansion of $(1+x)^{10}$.

$$T_{q+1} = {}_{10}C_q x^q \qquad (q = 0, 1, 2, \ldots, 10)$$

Let $T(p, q)$ represent the general term of the binomial $(1-x^3)(1+x)^{10}$ and then

$$T(p,q) = T_{p+1} \cdot T_{q+1} = (-1)^p \cdot {}_1C_p \cdot {}_{10}C_q \cdot x^{3p+q}.$$

Let $\qquad 3p + q = 5 \qquad (0 \leqslant p \leqslant 1), (0 \leqslant q \leqslant 10).$ (a)

To meet the condition (a) we have two eligible pairs of (p, q).

$$(0, 5), (1, 2).$$

It means there are two terms having x^5, $T(0, 5)$ and $T(1, 2)$, in the binomial expansion before combining like terms. Then the term x^5 in the binomial expansion after combining like terms becomes.

$$T(0,5) + T(1,2) = T_{0+1} \cdot T_{5+1} + T_{1+1} \cdot T_{2+1}.$$

The coefficient of the term x^5 becomes

$${}_1C_0 \cdot {}_{10}C_5 - {}_1C_1 \cdot {}_{10}C_2 = 207.$$

Example 2.1.7 The Term $x^s y^t$

Find the coefficient of the term x^3 in the expansion $(1+x)^4(2-x)^5$.

Solution:

We discuss the expression $(1+x)^4(2-x)^5$ in two separate parts.

1) Let T_{p+1} be the general term of the binomial expansion of $(1+x)^4$

$$T_{p+1} = {}_4C_p x^p \qquad (0 \leqslant p \leqslant 4).$$

2) Let T_{q+1} be the general term of the binomial expansion of $(2-x)^5$

$$T_{q+1} = {}_5C_q 2^{5-q}(-1)^q x^q \qquad (0 \leqslant q \leqslant 5).$$

Let $T(p, q)$ represent the general term of the binomial expansion of the expression

$$T(p,q) = T_{p+1} \cdot T_{q+1} = (-1)^q \cdot 2^{5-q} \cdot {}_4C_p \cdot {}_5C_q \cdot x^{p+q} \qquad (0 \leqslant p \leqslant 4, 0 \leqslant q \leqslant 5).$$

Let $\qquad p+q=3 \qquad (0 \leqslant p \leqslant 4), (0 \leqslant q \leqslant 5).$

To meet this condition we have four eligible pairs of (p, q):

$$(0, 3), (1, 2), (2, 1), (3, 0)$$

The the term x^3 we are looking for becomes

$$T(0,3) + T(1,2) + T(2,1) + T(3,0).$$

The coefficient of the term x^3 is

$$(-1)^3 \cdot 2^2 \cdot {}_4C_0 \cdot {}_5C_3 + (-1)^2 \cdot 2^3 \cdot {}_4C_1 \cdot {}_5C_2 + (-1)^1 \cdot 2^4 \cdot {}_4C_2 \cdot {}_5C_1 + (-1)^0 \cdot 2^5 \cdot {}_4C_3 \cdot {}_5C_0 = -72.$$

Example 2.1.8 The Term $x^s y^t$

Find the coefficient of the term x^4 in the expansion of the expression

$$(1+x) + (1+x)^2 + \cdots + (1+x)^{20}.$$

Solution:

Because there is no term of x^4 in the expansions $(1+x)$, $(1+x)^2$, and $(1+x)^3$, we will discuss the expressions for the following binomials

$$(1+x)^4 + (1+x)^5 + \cdots + (1+x)^{20}.$$

Then we have the coefficient for the term x^4 in final expression

$${}_4C_4 + {}_5C_4 + \cdots + {}_{20}C_4.$$

Since ${}_{n+1}C_m = {}_nC_m + {}_nC_{m-1}$, ${}_4C_4 = {}_5C_5$,

$$\begin{aligned}
{}_4C_4 + {}_5C_4 + {}_6C_4 + \cdots + {}_{20}C_4 &= {}_5C_5 + {}_5C_4 + {}_6C_4 + \cdots + {}_{20}C_4 \\
&= {}_6C_5 + {}_6C_4 + {}_7C_4 + \cdots + {}_{20}C_4 \\
&= {}_7C_5 + {}_7C_4 + {}_8C_4 + \cdots + {}_{20}C_4 \\
&\cdots \\
&= {}_{21}C_5.
\end{aligned}$$

Therefore the coefficient of the term x^4 in the expansion is ${}_{21}C_5 = 20349$.

2.1 Finding a Particular Term in a Binomial Expansion

Example 2.1.9   The Term $x^s y^t$

In the binomial expansion of the expression $(x-y+2z)^5$, find
1) the coefficient of the term $x^2 y z^2$
2) the number of all terms in the binomial expansion before combining like terms.

Solution:

1) Write $(x-y+2z)^5$ to $((x-y)+2z)^5$ then the general term of the expansion is

$$T_{k+1} = {}_5C_k (x-y)^{5-k}(2z)^k. \qquad (0 \leqslant k \leqslant 5)$$

Let T_{p+1} be the general term of the binomial $(x-y)^{5-k}$ in the formula above, then

$$T_{p+1} = {}_{5-k}C_p x^{5-k-p}(-y)^p. \qquad (0 \leqslant p \leqslant 3).$$

Let $T(k, p)$ replace T_{p+1}.

$$T(k, p) = T_{k+1} = {}_5C_k (x-y)^{5-k}(2z)^k$$
$$= {}_5C_k \cdot T_{p+1} \cdot (2z)^k$$
$$= -2^k \cdot {}_5C_k \cdot {}_{5-k}C_p \cdot x^{5-k-p} \cdot y^p \cdot z^k.$$

Let $k = 2$ and $p = 1$.

$$T(2,1) = -4 \cdot {}_5C_2 \cdot {}_3C_1 x^2 \cdot y \cdot z^2$$

The coefficient of the term $x^2 y z^2$ is $-4 \cdot {}_5C_2 \cdot {}_3C_1 = -120$.

2) There are six terms in the expansion of the binomial $(x-y+2z)^5$, which represented by $T(k, p)$. Because there are $5-k+1$ terms in the binomial $(x-y)^{5-k}$ in $T(k, p)$, then

$$\sum_{k=0}^{5} 5-k+1 = 21.$$

There are *21* terms in the binomial expansion of the binomial $(x-y+2z)^5$.

Example 2.1.10   The Term $x^s y^t$

Find the coefficient of the term x^5 in the expansion of the expression $(x^2 - 2x + 2)^6$.

Solution:

Rearrange the expression to the binomial

$$(x^2 - 2x + 2)^6 = (1 + (x-1)^2)^6.$$

The the general term of expansion of the binomial $(1+(x-1)^2)^6$ is

$$T_{k+1} = {}_6C_k (x-1)^{2k} \qquad (0 \leqslant k \leqslant 6).$$

Let the general term of expansion of the binomial $(x-1)^{2k}$ within T_{k+1} be
$$T_{p+1} = {}_{2k}C_p x^{2k-p} \cdot (-1)^p \qquad (0 \leq p \leq 2k)$$
Let $T(k, p)$ replace T_{k+1} then
$$T(k, p) = T_{k+1} = {}_6C_k \cdot T_{p+1} = (-1)^p \cdot {}_6C_k \cdot {}_{2k}C_p \cdot x^{2k-p}$$
Let $2k - p = 5$ and we have following conditions
$$\begin{cases} 0 \leq k \leq 6 \\ 0 \leq p \leq 2k \\ 2k - p = 5 \end{cases}$$
Four pairs (k, p) meet the conditions $(3, 1)$, $(4, 3)$, $(5, 5)$, and $(6, 7)$. In other words there are four terms having x^5. After combining like terms the coefficient of the term x^5 becomes
$$T(3,1) + T(4,3) + T(5,5) + T(6,7)$$
$$= -{}_6C_3 \cdot {}_6C_1 - {}_6C_4 \cdot {}_8C_3 - {}_6C_5 \cdot {}_{10}C_5 - {}_6C_6 \cdot {}_{12}C_7 = -3{,}264.$$

Example 2.1.11   The Term $x^s y^t$

Find the coefficient of the term x^5 in the expansion of the expression $(2 + 3x + 4x^2)^8$.

Solution:

Arrange the expression to a binomial
$$(2 + 3x + 4x^2)^8 = ((2 + 3x) + 4x^2)^8.$$
The general term of the expansion of the binomial $((2+3x) + 4x^2)^8$ is
$$T_{k+1} = {}_8C_k (2+3x)^{8-k} \cdot (4x^2)^k \qquad (0 \leq k \leq 8).$$
For the binomial $(2+3x)^{8-k}$ within T_{k+1}, let its general term be
$$T_{p+1} = {}_{8-k}C_p 2^{8-k-p} \cdot (3x)^p \qquad (0 \leq p \leq 8-k)$$
Let $T(k, p)$ replace T_{k+1} then
$$T(k, p) = T_{k+1} = {}_8C_k \cdot T_{p+1} \cdot (4x^2)^k = 2^{8-k-p} \cdot 3^p \cdot 4^k \cdot {}_8C_k \cdot {}_{8-k}C_p \cdot x^{2k+p}$$
Let $2k + p = 5$ and we have following conditions
$$\begin{cases} 0 \leq k \leq 8 \\ 0 \leq p \leq 8-k \\ 2k + p = 5 \end{cases}$$
There are three pairs $(0, 5)$, $(1, 3)$ and $(2, 1)$, to meet the conditions above. In other words there are three terms of x^5 before combining like terms. After combining like terms the coefficient of the term x^5 will be
$$T(0,5) + T(1,3) + T(2,1)$$
$$= 2^3 \cdot 3^5 \cdot 4^0 \cdot {}_8C_0 \cdot {}_8C_5 + 2^4 \cdot 3^3 \cdot 4 \cdot {}_8C_1 \cdot {}_7C_3 + 2^5 \cdot 3 \cdot 4^2 \cdot {}_8C_2 \cdot {}_6C_1 = 850{,}752.$$

2.1 Finding a Particular Term in a Binomial Expansion

▶ Constant Term

To find constant term(s) we can set the power of the variable(s) of the general term of the binomial expansion to zero. There may have more than one constant terms in the binomial expansion before combining like terms and one constant term is obtained after like terms combined.

Example 2.1.12 Constant Term

Find constant term(s) of the expansion of the expression $\left(x+\dfrac{1}{x}\right)^4 (x^2+1)^6$.

Solution:

Let T_{p+1} be the general term of the binomial expansion of the binomial $\left(x+\dfrac{1}{x}\right)^4$.

$$T_{p+1} = {}_4C_p\, x^{4-p}\left(\dfrac{1}{x}\right)^p = {}_4C_p\, x^{4-2p} \qquad (0 \leqslant p \leqslant 4)$$

Let T_{q+1} be the general term of the binomial expansion of the binomial $(x^2+1)^6$.

$$T_{q+1} = {}_6C_q (x^2)^{6-q} = {}_6C_q\, x^{12-2q} \qquad (0 \leqslant q \leqslant 6).$$

Let $T(p, q)$ be the general term of the binomial expansion of $\left(x+\dfrac{1}{x}\right)^4 (x^2+1)^6$.

$$T(p, q) = T_{p+1} \cdot T_{q+1} = ({}_4C_p\, x^{4-2p})({}_6C_q\, x^{12-2q})$$
$$= {}_4C_p \cdot {}_6C_q\, x^{16-2p-2q} \qquad (0 \leqslant p \leqslant 4,\, 0 \leqslant q \leqslant 6)$$

To make $T(p, q)$ a constant, let $16 + 2p - 2q = 0$ and we have the following conditions.

$$\begin{cases} 0 \leqslant p \leqslant 4 \\ 0 \leqslant q \leqslant 6 \\ 16 - 2p - 2q = 0 \end{cases}$$

We list all pairs of (p, q) which meet above conditions:

$$(2, 6),\ (3, 5),\ (4, 4).$$

There are three constant terms in the binomial expansion before combining like terms.

$$T(2, 6) = {}_4C_2 \cdot {}_6C_6; \qquad T(3, 5) = {}_4C_3 \cdot {}_6C_5; \qquad T(4, 4) = {}_4C_4 \cdot {}_6C_4;$$

After combining like terms the constant term in the binomial expansion becomes

$$T(2, 6) + T(3, 5) + T(4, 4) = {}_4C_2 \cdot {}_6C_6 + {}_4C_3 \cdot {}_6C_5 + {}_4C_4 \cdot {}_6C_4$$
$$= 45.$$

Example 2.1.13   Constant Term

Find constant term(s) in the binomial expansion of the binomial
$$(\frac{3}{\sqrt{x}}-2\sqrt{x})^6 \ (x>0).$$

Solution:

As
$$(\frac{3}{\sqrt{x}}-2\sqrt{x})^6 = \frac{(3-2x)^6}{x^3},$$

the binomial expansion of $\frac{(3-2x)^6}{x^3}$ may have a constant term if there exists a term of x^3 in the binomial expansion of $(3-2x)^6$.

Let T'_{k+1} be the general term of the binomial expansion of $(3-2x)^6$.
$$T'_{k+1} = {}_6C_k 3^{6-k}(-2x)^k$$

Let T_{k+1} be the general term of the binomial expansion of $\frac{(3-2x)^6}{x^3}$ then

$$T_{k+1} = \frac{T'_{k+1}}{x^3} = \frac{{}_6C_k 3^{6-k}(-2x)^k}{x^3} = {}_6C_k 3^{6-k}(-2)^k \cdot x^{k-3}.$$

To make T_{k+1} a constant term, let $k = 3$ and the fourth term of the binomial expansion becomes a constant term.
$$T_{3+1} = {}_6C_3 3^3 (-2)^3 = -4,320.$$

Example 2.1.14   Constant Term

In the binomial expansion of the binomial $(x^3 - \frac{1}{x})^n$ $(x \neq 0)$, the constant term is -252, find $n = ?$

Solution:

$$T_{k+1} = {}_nC_k (x^3)^{n-k}(-\frac{1}{x})^k = (-1)^k {}_nC_k x^{2n-4k}.$$

To make T_{k+1} a constant term the following conditions should be met according to the facts given.

(a) Let $2n - 4k = 0$ and $k = \frac{n}{2}$ $(0 \leqslant k \leqslant n)$.

(b) Because the constant term is negative, k should be a odd number.

(c) $T_{k+1} = -252$

2.1 Finding a Particular Term in a Binomial Expansion

Then we have the following potential pairs of (n, k).

(n, k)	Constant Terms
$(2, 1)$	$T_2 = -{}_2C_1 = -2$
$(6, 3)$	$T_4 = -{}_6C_3 = -20$
$(10, 5)$	$T_6 = -{}_{10}C_5 = -252$
...	...

As $T_6 = -252$, we obtain $n = 10$.

Example 2.1.15   Constant Term

Find constant term(s) in the binomial expansion of the binomial
$$\left(2x - \frac{1}{\sqrt{x}}\right)^n \quad (x > 0)$$

Solution:

We can write the general term of the binomial expansion as below.

$$T_{k+1} = {}_nC_k (2x)^{n-k} \left(-\frac{1}{\sqrt{x}}\right)^k = (-1)^k 2^{n-k} {}_nC_k x^{n-\frac{3}{2}k}$$

To make the general term T_{k+1} a constant term, let

$$n - \frac{3}{2}k = 0.$$

$$k = \frac{2}{3}n.$$

As both k and n are integers, n should be a multiple of 3. We have the following pairs of (n, k).

(n, k)	Constant Terms
$(3, 2)$	$T_{2+1} = (-1)^2 \cdot 2 \cdot {}_3C_2 = 3$
$(6, 4)$	$T_{4+1} = (-1)^4 \cdot 2^2 \cdot {}_6C_4 = 60$
$(9, 6)$	$T_{6+1} = (-1)^6 \cdot 2^3 \cdot {}_9C_6 = 672$
...	...

When the power n is known the constant term is determined.

Example 2.1.16 Constant Term

Find constant term(s) in the binomial expansion of the expression $\left(1+x+\dfrac{1}{x}\right)^8$ $(x \neq 0)$.

Solution:

As
$$\left(1+x+\dfrac{1}{x}\right)^8 = \left(1+\left(x+\dfrac{1}{x}\right)\right)^8,$$

the general term of the binomial expansion becomes

$$T_{k+1} = {}_8C_k \left(x+\dfrac{1}{x}\right)^k \qquad (k = 0, 1, 2, \ldots, 8).$$

Let T'_{p+1} be the general term of the binomial expansion of $\left(x+\dfrac{1}{x}\right)^k$.

$$T'_{p+1} = {}_kC_p\, x^{k-p}\left(\dfrac{1}{x}\right)^p = {}_kC_p\, x^{k-2p} \qquad (0 \leq p \leq k).$$

Let $T(k, p)$ represent T_{k+1} and we have

$$T(k, p) = {}_8C_k \cdot T'_{p+1} = {}_kC_p\, x^{k-2p}.$$

To make $T(k, p)$ a constant, let $k - 2p = 0$, we have $2p = k$. Therefore constant terms emerge when k and p show in the following pairs (k, p).

$(0, 0), \quad (2, 1), \quad (4, 2), \quad (6, 3), \quad (8, 4)$

There are five constant terms in the binomial expansion before combining like terms.

$$\begin{cases} T(0,0) = {}_8C_0 \cdot {}_0C_0 \\ T(2,1) = {}_8C_2 \cdot {}_2C_1 \\ T(4,2) = {}_8C_4 \cdot {}_4C_2 \\ T(6,3) = {}_8C_6 \cdot {}_6C_3 \\ T(8,4) = {}_8C_8 \cdot {}_8C_4 \end{cases}$$

After combining like terms, the constant term in the binomial expansion becomes

$$T(0,0) + T(2,1) + \cdots + T(8,4)$$
$$= {}_8C_0 \cdot {}_0C_0 + {}_8C_2 \cdot {}_2C_1 + {}_8C_4 \cdot {}_4C_2 + {}_8C_6 \cdot {}_6C_3 + {}_8C_8 \cdot {}_8C_4$$
$$= 1{,}107.$$

2.1 Finding a Particular Term in a Binomial Expansion

▶ Rational Coefficients and Terms

We give the definition of rational coefficient and rational term as below.

Rational Coefficient   A coefficient whose power is an integer.
Rational Term   A term whose coefficient is rational and the powers of all variables of the term are integers.

The methods below are used to find rational coefficient(s) or rational term(s) in a binomial expansion.

- To find **rational coefficient(s)**, set the power of the coefficient of the general term to an integer.
- To find **rational term(s)**, set the power of the coefficient and all variables of the general term to integers.

Example 2.1.17   *Rational Coefficient and Term*

In the binomial expansion of the binomial $(\sqrt{2}+\sqrt[3]{3}\sqrt[4]{x})^{20}$,
1) how many terms whose coefficient is rational?
2) how many rational terms?

Solution:

1) The general term of the binomial expansion of $(\sqrt{2}+\sqrt[3]{3}\sqrt[4]{x})^{20}$ is

$$T_{k+1} = {}_{20}C_k (\sqrt{2})^{20-k} (\sqrt[3]{3}\sqrt[4]{x})^k = {}_{20}C_k \, 2^{10-\frac{k}{2}} \, 3^{\frac{k}{3}} \, x^{\frac{k}{4}} \quad (0 \leqslant k \leqslant 20)$$

Let t_{k+1} be the coefficient of term T_{k+1} and it becomes

$$t_{k+1} = {}_{20}C_k \, 2^{10-\frac{k}{2}} \, 3^{\frac{k}{3}} \quad (0 \leqslant k \leqslant 20)$$

To make t_{k+1} rational the two conditions below should be met.

- $\dfrac{k}{3}$ should be an integer then k must be a multiple of 3. Within the range of $0 \leqslant k \leqslant 20$, we have these potential values underlined
$$k = \{\, \underline{0},\ 3,\ \underline{6},\ 9,\ \underline{12},\ 15,\ \underline{18}\,\}.$$

- $10 - \dfrac{k}{2}$ should also be an integer and we have these potential values underlined as below
$$k = \{\, \underline{0},\ 2,\ 4,\ \underline{6},\ 8,\ 10,\ \underline{12},\ 14,\ 16,\ \underline{18},\ 20\,\}$$

From two lists above we pick all numbers in common from two lists,
$$k = \{ 0, 6, 12, 18 \}. \tag{a}$$

Thus we obtain four terms T_1, T_7, T_{13}, and T_{19} that have rational binomial coefficients.

2) To make $T_{k+1} = t_{n+1} \cdot x^{\frac{k}{4}}$ rational, the power of variable x, $\frac{k}{4}$, should be an integer and k a multiple of 4. In the list (a), $k = 12$ is the eligible value only. Then there is one rational term in the binomial expansion.
$$T_{13} = t_{13} \cdot x^3 = 5{,}184 \cdot {}_{20}C_{12} \cdot x^3$$

Example 2.1.18   Rational Coefficient and Term

Find all rational term(s) in the binomial expansion of the binomial
$$(\sqrt[3]{2} + \frac{1}{\sqrt{3}})^{18}.$$

Solution:

Let the general term of the binomial expansion be
$$T_{k+1} = {}_{18}C_k (\sqrt[3]{2})^{18-k} (3^{-\frac{1}{2}})^k = {}_{18}C_k \, 2^{6-\frac{k}{3}} \cdot 3^{-\frac{k}{2}} \quad (0 \leq k \leq 18).$$

- To make $6 - \frac{k}{3}$ an integer, k should be a multiple of 3. Then potential values for k are
$$k = \{ 0, 3, 6, 9, \underline{12}, 15, \underline{18} \}.$$

- To make $-\frac{k}{2}$ an integer, k should be a multiple of 2. Then potential values for k are
$$k = \{ 0, 2, 4, 8, 10, \underline{12}, 14, 16, \underline{18} \}.$$

The values of k which are common in both lists are 12 and 18. Then we obtain two rational terms, the thirteen and the nineteen term of the binomial expansion.
$$T_{13} = {}_{18}C_{12} \, 2^2 \cdot 3^{-6} = \frac{74{,}256}{729}$$
$$T_{19} = {}_{18}C_{18} \cdot 3^{-9} = \frac{1}{19{,}683}.$$

2.1 Finding a Particular Term in a Binomial Expansion

Example 2.1.19   Rational Coefficient and Term

In the binomial expansion of the binomial $(2\sqrt[3]{x}+\sqrt[4]{5})^{24}$, find
1) how many terms with rational coefficient?
2) how many rational terms?

Solution:

The general term of the binomial is

$$T_{k+1} = {}_{24}C_k (2\sqrt[3]{x})^{24-k} \cdot (\sqrt[4]{5})^k = {}_{24}C_k \cdot (2^{k-24} \cdot 5^{\frac{k}{4}}) \cdot x^{\frac{24-k}{3}} \quad (0 \leqslant k \leqslant 24).$$

1) To have the coefficient, ${}_{24}C_k \cdot (2^{24-k} \cdot 5^{\frac{k}{4}})$, a rational, the power $\dfrac{k}{4}$ should be an integer and k should be a multiple of 4. We have the following potential values for k.

$$k = \{\, 0,\ 4,\ 8,\ \underline{12},\ 16,\ 20,\ \underline{24}\,\} \qquad\qquad (a)$$

Then there are seven terms whose coefficients are rational and they are the 1st, 5th, 9th, 13th, 17th, 21st, and the 25th term.

2) Because a rational term is the term whose both coefficient and variables are rational, the power $\dfrac{24-k}{3}$ of variable x should be an integer and k should be a multiple of 3. We have the following potential values for k.

$$k = \{\, 0,\ 3,\ 6,\ 9,\ \underline{12},\ 15,\ 18,\ 21,\ \underline{24}\,\} \qquad\qquad (b)$$

From two lists (a) and (b) we found that two values *12* and *24* are in common.

Therefore there are two rational terms, the 13th term and the 25th term, i.e.

$$T_{13} = {}_{24}C_{12} \cdot (2^{12} \cdot 5^3) \cdot x^4$$
$$T_{25} = 5^3$$

in the binomial expansion of the binomial.

▶ The Term with the Maximum Binomial Coefficient

To find the maximum binomial coefficient in a binomial expansion we should know that the maximum binomial coefficient(s) is a number of combination, which is positive, and associated with the middle term(s) of the binomial expansion. The maximum binomial coefficient(s) in the binomial expansion depends on the power n of the binomial only. Thus we don't have to expand the binomial to find them.

1) If the power n of the binomial is an even number there is a single middle term
   - The $(\frac{n}{2}+1)^{th}$ term is the middle term.
   - Its binomial coefficient, $_nC_{\frac{n}{2}}$, is the maximum binomial coefficient.

2) If the power n of the binomial is a odd number there are two middle terms.
   - Both the $(\frac{n+1}{2})^{th}$ term and the $(\frac{n+1}{2}+1)^{th}$ term are the middle terms of the binomial expansion.
   - Their binomial coefficients are the maximum binomial coefficients which are the same.
   $$_nC_{\frac{n-1}{2}} = {_nC_{\frac{n+1}{2}}}.$$

Example 2.1.20   The Term with the Maximum Binomial Coefficient

Find the maximum binomial coefficient in the binomial expansion of the following binomials.

1) $(x+y)^4$ 2) $(2x+3y)^5$ 3) $(5x-2y)^{17}$ 4) $(ax-by)^{2m-1}$ $(a, b > 0)$

Solution:

1) The binomial coefficients are: *1 4 **6** 4 1*.
   As $n = 4$ is an even number, the middle term is the 3th term and its binomial coefficient
   $$_4C_{\frac{4}{2}} = 6$$
   is the maximum binomial coefficient in the binomial expansion.

2) The binomial coefficients are: *1 5 **10 10** 5 1*.
   As $n = 5$ is an even number, there are two middle terms, the 3th term and the 4th term. Their binomial coefficients,
   $$_5C_{\frac{5-1}{2}} = {_5C_2} = 10$$

2.1 Finding a Particular Term in a Binomial Expansion

and
$$_5C_{\frac{5+1}{2}} = {_5C_3} = 10,$$

are the maximum binomial coefficient in the binomial expansion.

3) As $n = 17$ is a odd number, both the $(\frac{17+1}{2})^{th}$ term and the $(\frac{17+1}{2}+1)^{th}$ term, i.e. the ninth term and the tenth term, have the maximum binomial coefficient

$$_{17}C_{\frac{17-1}{2}} = {_{17}C_8} = 24{,}310$$

and
$$_{17}C_{\frac{17+1}{2}} = {_{17}C_9} = 24{,}310.$$

4) Since $2m - 1$ is a odd number, there are two middle terms which are the mth term and the $(m+1)^{th}$ term. Their binomial coefficients, $_{2m-1}C_{m-1}$ and $_{2m-1}C_m$ are the maximum binomial coefficient.

Example 2.1.21 The Term with the Maximum Binomial Coefficient

In the binomial expansion of the binomial $(3+\frac{x}{3})^n$, the binomial coefficients of the 5th term, the 6th term, and the 7th term consist of a arithmetic sequence. Find the maximum binomial coefficient in the binomial expansion.

Solution:

The binomial coefficients of the 5th term, the 6th term, and the 7th term are

$$_nC_4, \ _nC_5, \ _nC_6.$$

By arithmetic mean, we have

$$_nC_4 + {_nC_6} = 2 \ _nC_5.$$

That is
$$\frac{n!}{4!(n-4)!} + \frac{n!}{6!(n-6)!} = 2\frac{n!}{5!(n-5)!}.$$

After solving the above equation we obtain two solutions $n = 7$ and $n = 14$.

- When $n = 7$, the binomial $(3+\frac{x}{3})^7$ has 8 terms in its binomial expansion. There are two middle terms, the 4th term and 5th term. The maximum binomial coefficient in the binomial expansion is

$$_7C_3 = {_7C_4} = 35.$$

- When $n = 14$, the binomial $(3+\frac{x}{3})^{14}$ has 15 terms in its binomial expansion. There is a single middle term, the 8th term. The maximum binomial coefficient in the binomial expansion is

$$_{14}C_8 = 3{,}432.$$

Example 2.1.22   The Term with the Maximum Binomial Coefficient

In the binomial expansion of a binomial $\left(x+\dfrac{1}{x}\right)^n$, the sum of all binomial coefficients in the binomial expansion is 256. Find the maximum binomial coefficient in the binomial expansion.

Solution:

By the property of the binomial coefficients, we have the sum of all binomial coefficients of a binomial expansion as below.

$$_nC_0 + {}_nC_1 + \cdots + {}_nC_k + \cdots + {}_nC_{n-1} + {}_nC_n = 2^n$$

Let $2^n = 256$ and we obtain $n = 8$. There are total 9 terms in the binomial expansion. Thus there is one middle term, the fifth term T_5, and its coefficient

$$_8C_4 = 70.$$

is the maximum binomial coefficient in the binomial expansion.

▶ The Term with the Maximum Coefficient

> **DEFINITION   The Maximum Coefficient in a Binomial Expansion**
>
> In a binomial expansion, if at least one positive coefficient exists the maximum coefficient is the coefficient with the maximum absolute value in all positive coefficients else the maximum coefficient is that with the minimum absolute value in all negative coefficients.

Unlike the maximum binomial coefficient, the maximum coefficient in a binomial expansion has the following features.
- It may be negative.
- It does not necessarily associate with the term with the maximum binomial coefficient. Take $(ax+by)^n$, $(a,b \in R)$, as an example. a and b are the coefficients of variable x and y respectively in the binomial. The position of the term with the maximum coefficient in the binomial expansion depends on the sign and value of a and b. Suppose that $a > 0$ and $b > 0$. If $a > b$, the position of the term having the maximum coefficient may be located to the left from the middle of the binomial expansion else to the right.

We give a general guide to identify the maximum coefficient as below.
1) Look for the term whose coefficient has the maximum absolute value in all

2.1 Finding a Particular Term in a Binomial Expansion

coefficients of the binomial expansion. Let the general term of $(x+y)^n$,
$$T_{k+1} = {}_nC_k x^{n-k} y^k \quad (n, k \in N^+, 0 \leq k \leq n),$$
be the term whose coefficient has the maximum absolute value in all coefficients. Let t_k, t_{k+1}, and t_{k+2} be the coefficient of terms T_k, T_{k+1}, and T_{k+2} respectively.

2) Let the following inequality set hold.
$$\begin{cases} |t_{k+1}| \geq |t_k| \\ |t_{k+1}| \geq |t_{k+2}| \end{cases}$$

After solving above inequalities we obtain one or two solutions for k. Take k as positive integer. Then the coefficient of the $(k+1)^{th}$ term has the maximum absolute value. If the coefficient is positive it is the maximum coefficient we are looking for else it is the minimum coefficient and the maximum coefficient can be found between the $(k+1)^{th}$ term's two neighbors.

The following examples show you how to identify the maximum coefficient in a binomial expansion. For your convenience the indicators as below are used to show the middle term and different type of coefficients respectively in a binomial expansion.

✛	the middle term(s)
△	the maximum binomial coefficient
▲	the maximum coefficient
■	the coefficient having the maximum absolute value

Example 2.1.23 The Term with the Maximum Coefficient

In the binomial expansion of the binomial $(7x+2y)^4$ find
1) the maximum binomial coefficient
2) the term whose coefficient has the maximum absolute value
3) the maximum coefficient

Solution:

1) As $n = 4$ is an even number there is one middle term, T_3, with the maximum binomial coefficient in the binomial expansion
$${}_4C_2 = 6.$$

2) The general term of the binomial expansion is
$$T_{k+1} = {}_4C_k (7x)^{4-k} (2y)^k \quad (0 \leq k \leq 4).$$

Let term T_{k+1} be the term whose coefficient has the maximum absolute value in the binomial expansion. Its two neighbors are T_k and T_{k+2}.

$$\begin{cases} T_k = {}_4C_{k-1}(7x)^{4-k+1}(2y)^{k-1} \\ T_{k+2} = {}_4C_{k+1}(7x)^{4-k-1}(2y)^{k+1} \end{cases}$$

Let t_k, t_{k+1}, and t_{k+2} be the coefficient of terms T_{k+1}, T_k, and T_{k+2} respectively.

$$\begin{cases} t_{k+1} = {}_4C_k \cdot 7^{4-k} \cdot 2^k \\ t_k = {}_4C_{k-1} \cdot 7^{4-k+1} \cdot 2^{k-1} \\ t_{k+2} = {}_4C_{k+1} \cdot 7^{4-k-1} \cdot 2^{k+1} \end{cases}$$

Let the following inequalities hold.

$$\begin{cases} |t_{k+1}| \geq |t_k| \\ |t_{k+1}| \geq |t_{k+2}| \end{cases}$$

Then

$$\begin{cases} |{}_4C_k \cdot 7^{4-k} \cdot 2^k| \geq |{}_4C_{k-1} \cdot 7^{4-k+1} \cdot 2^{k-1}| \\ |{}_4C_k \cdot 7^{4-k} \cdot 2^k| \geq |{}_4C_{k+2} \cdot 7^{4-k-1} \cdot 2^{k+1}| \end{cases}$$

We obtain $\dfrac{1}{9} \leq k \leq \dfrac{10}{9}$ and take $k = 1$. Then $|t_2|$ is the maximum absolute value.

3) From 2) we know that the second term

$$T_2 = {}_4C_1(7x)^3(2y)$$

has the maximum absolute value in the binomial expansion and it's coefficient is

$$t_2 = {}_4C_1 \cdot 7^3 \cdot 2 = 2,744.$$

Since $t_2 > 0$, it is the maximum coefficient in the binomial expansion.

We expand the binomial to verify above results.

$(7x+2y)^4$
$= {}_4C_0(7x)^4 + {}_4C_1 \cdot (7x)^3(2y) + {}_4C_2 \cdot (7x)^2(2y)^2 + {}_4C_1 \cdot (7x)(2y)^3 + {}_4C_4(2y)^4$
△ +
$= 2,401\, x^5 + 2,744\, x^4 y + 1,176\, x^2 y^2 + 224\, x\, y^3 + 16\, y^4$
▲ ■   △ +

The second term T_2 has the maximum coefficient $2,744$ in the binomial expansion.

2.1 Finding a Particular Term in a Binomial Expansion

Example 2.1.24 *The Term with the Maximum Coefficient*

In the binomial expansion of the binomial $(2x-7y)^4$ find
1) the maximum binomial coefficient
2) the term whose coefficient is the maximum absolute value
3) the maximum coefficient

Solution:

1) As $n = 4$ is an even number there is one middle term, T_3, with the maximum binomial coefficient in the binomial expansion,

$$_4C_2 = 6.$$

2) The general term of the binomial $(2x-7y)^4$ is

$$T_{k+1} = {}_4C_k(2x)^{4-k}(-7y)^k \quad (0 \leqslant k \leqslant 4).$$

Let term T_{k+1} be the term whose coefficient has the maximum absolute value in the expansion. And we have its two neighbors T_k and T_{k+2} as well.

$$\begin{cases} T_k = {}_4C_{k-1}(2x)^{4-k+1}(-7y)^{k-1} \\ T_{k+2} = {}_4C_{k+2}(2x)^{4-k-1}(-7y)^{k+1} \end{cases}$$

Let t_k, t_{k+1}, and t_{k+2} be the coefficient of terms T_{k+1}, T_k, and T_{k+2} respectively.

$$\begin{cases} t_{k+1} = {}_4C_k \cdot 2^{4-k} \cdot (-7)^k \\ t_k = {}_4C_{k-1} \cdot 2^{4-k+1} \cdot (-7)^{k-1} \\ t_{k+2} = {}_4C_{k+1} \cdot 2^{4-k-1} \cdot (-7)^{k+1} \end{cases}$$

Let the following inequalities hold

$$\begin{cases} |t_{k+1}| \geqslant |t_k| \\ |t_{k+1}| \geqslant |t_{k+2}| \end{cases}$$

That is

$$\begin{cases} |{}_4C_k \cdot 2^{4-k} \cdot 7^k| \geqslant |{}_4C_{k-1} \cdot 2^{4-k+1} \cdot 7^{k-1}| \\ |{}_4C_k \cdot 2^{4-k} \cdot 7^k| \geqslant |{}_4C_{k+1} \cdot 2^{4-k-1} \cdot 7^{k+1}| \end{cases}$$

We obtain $\dfrac{26}{9} \leqslant k \leqslant \dfrac{35}{9}$ and take $k = 3$. Then $|t_4|$ is the maximum absolute value in the binomial expansion.

3) From the above we know that the fourth term

$$T_4 = {}_4C_3(2x)(-7y)^3$$

has the maximum absolute value in the binomial expansion and it's coefficient is

$$t_4 = {}_4C_3 \cdot 2 \cdot (-7)^3 = -2{,}744.$$

As $t_4 < 0$ it is the minimum coefficient in the binomial expansion. The maximum coefficient can be one of its neighbors and let's check $t_3 = 1{,}176$ and $t_5 = 2{,}401$. Because $t_3, t_5 > 0$ and $t_5 > t_3$ then the coefficient of the fifth term, t_5, is the maximum coefficient in the binomial expansion.

Now we expand the binomial to verify above results.

$(2x-7y)^4$
$= {}_4C_0(2x)^4 + {}_4C_1 \cdot (2x)^3(-7y) + {}_4C_2 \cdot (2x)^2(-7y)^2 + {}_4C_3 \cdot (2x)(-7y)^3 + {}_4C_4(-7y)^4$
$\quad\quad\quad\quad\quad\quad\quad\quad\quad\quad\quad\quad\quad \triangle + $
$= 16x^4 - 224x^3y + 1{,}176x^2y^2 - 2{,}744xy^3 + 2{,}401y^4$
$\quad\quad \triangle + \quad\quad\quad \blacksquare \quad\quad\quad\quad \blacktriangle$

The fifth term T_5 has the maximum coefficient, $2{,}401$, in the binomial expansion.

Example 2.1.25   The Term with the Maximum Coefficient

In the binomial expansion of $(7x - 2y)^5$ find
1) the maximum binomial coefficient
2) the term whose coefficient is the maximum absolute value
3) the maximum coefficient

Solution:

1) As $n = 5$ is a odd number there are two middle terms, T_3 and T_4, with the maximum binomial coefficients,

$${}_5C_2 = {}_5C_3 = 10.$$

2) The general term of the binomial $(7x - 2y)^5$ is

$$T_{k+1} = {}_5C_k(7x)^{5-k}(-2y)^k \quad (0 \leq k \leq 5).$$

Let T_{k+1} be the term whose coefficient has the maximum absolute value in the expansion. And we have its two neighbors T_k and T_{k+2} as well.

$$\begin{cases} T_k = {}_5C_{k-1}(7x)^{5-k+1}(-2y)^{k-1} \\ T_{k+2} = {}_5C_{k+1}(7x)^{5-k-1}(-2y)^{k+1} \end{cases}$$

Let t_k, t_{k+1}, and t_{k+2} be the coefficient of terms T_{k+1}, T_k, and T_{k+2} respectively.

$$\begin{cases} t_{k+1} = {}_5C_k \cdot 7^{5-k} \cdot (-2)^k \\ t_k = {}_5C_{k-1} \cdot 7^{5-k+1} \cdot (-2)^{k-1} \\ t_{k+2} = {}_5C_{k+1} \cdot 7^{5-k-1} \cdot (-2)^{k+1} \end{cases}$$

2.1 Finding a Particular Term in a Binomial Expansion

Let the following inequalities hold.
$$\begin{cases} |t_{k+1}| \geq |t_k| \\ |t_{k+1}| \geq |t_{k+2}| \end{cases}$$
$$\begin{cases} {}_5C_k \cdot 7^{5-k} \cdot 2^k \geq {}_5C_{k-1} \cdot 7^{5-k+1} \cdot 2^{k-1} \\ {}_5C_k \cdot 7^{5-k} \cdot 2^k \geq {}_5C_{k+1} \cdot 7^{5-k-1} \cdot 2^{k+1} \end{cases}$$

We obtain $\frac{33}{9} \leq k \leq \frac{42}{9}$ and take $k = 1$. Then $|t_2|$ is the maximum absolute value in the binomial expansion.

3) From 2) we know that the second term
$$T_2 = {}_5C_1 (7x)^4 (-2y) = -24{,}010 \, x^4 y$$
has the maximum absolute value in the binomial expansion and it's coefficient is
$$t_2 = -24{,}010.$$

As $t_2 < 0$ it is the minimum coefficient in the binomial expansion. The maximum coefficient can be one of its neighbors and let's check $t_1 = 16{,}807$ and $t_3 = 13{,}720$. Because both are positive and $t_1 > t_3$ then t_1 is the maximum coefficient in the binomial expansion.

Now we expand the binomial to verify above results.
$(7x - 2y)^5$
$= (7x)^5 + 5 \cdot (7x)^4(-2y) + 10 \cdot (7x)^3(-2y)^2 + 10 \cdot (7x)^2(-2y)^3 + 5 \cdot (7x)(-2y)^4 + (-2y)^5$
$\qquad\qquad\qquad \Delta + \qquad\qquad\qquad\qquad \Delta +$
$= 16{,}807 \, x^5 - 24{,}010 \, x^4 y + 13{,}720 \, x^3 y^2 - 3{,}920 \, x^2 y^3 + 560 x \, y^4 - 32 \, y^5$
▲ ■ $\qquad\qquad \Delta + \qquad\qquad \Delta +$

The first term T_1 has the maximum coefficient $16{,}807$ in the binomial expansion.

Example 2.1.26   The Term with the Maximum Coefficient

In the binomial expansion of $\left(2x - \frac{1}{x}\right)^6$ find

1) the maximum binomial coefficient
2) the term whose coefficient is the maximum absolute value
3) the maximum coefficient

Solution:

1) As $n = 6$ is an even number there is one middle term, T_4. The maximum binomial coefficient in the binomial expansion is
$${}_6C_3 = 20.$$

2) The general term of the binomial is
$$T_{k+1} = {}_6C_k 2^{6-k}(-1)^k x^{6-2k} \quad (0 \leq k \leq 6).$$
Let term T_{k+1} be the term whose coefficient has the maximum absolute value in the expansion. And we have its two neighbors T_k and T_{k+2} as well.
$$\begin{cases} T_k = {}_6C_{k-1} 2^{6-k+1}(-1)^{k-1} x^{6-2(k-1)} \\ T_{k+2} = {}_6C_{k+1} 2^{6-k-1}(-1)^{k+1} x^{6-2(k+1)} \end{cases}$$
Let t_k, t_{k+1}, and t_{k+2} be the coefficient of terms T_{k+1}, T_k, and T_{k+2} respectively.
$$\begin{cases} t_{k+1} = {}_6C_k \cdot 2^{6-k} \cdot (-1)^k \\ t_k = {}_6C_{k-1} \cdot 2^{6-k+1} \cdot (-1)^{k-1} \\ t_{k+2} = {}_6C_{k+1} \cdot 2^{6-k-1} \cdot (-1)^{k+1} \end{cases}$$
Let the following inequalities hold.
$$\begin{cases} |t_{k+1}| \geq |t_k| \\ |t_{k+1}| \geq |t_{k+2}| \end{cases}$$
That is
$$\begin{cases} |{}_6C_k \cdot 2^{6-k}| \geq |{}_6C_{k-1} \cdot 2^{6-k+1}| \\ |{}_6C_k \cdot 2^{6-k}| \geq |{}_6C_{k+1} \cdot 2^{6-k-1}| \end{cases}$$

We obtain $\frac{4}{3} \leq k \leq \frac{7}{3}$ and take $k = 2$. Then $|t_3|$ is the maximum absolute value in the binomial expansion.

3) From 2) we know that the second term
$$T_3 = {}_6C_2 2^4 (-1)^2 x^2 = 240 x^2$$
has the maximum absolute value in the binomial expansion and it's coefficient is
$$t_3 = 240.$$
Since $t_3 > 0$, it is the maximum coefficient in the binomial expansion.

Now we expand the binomial to verify above results.
$$\left(2x - \frac{1}{x}\right)^6 =$$
$${}_6C_0(2x)^6 + {}_6C_1\frac{(2x)^5}{-x} + {}_6C_2\frac{(2x)^4}{(-x)^2} + {}_6C_3\frac{(2x)^3}{(-x)^3} + {}_6C_4\frac{(2x)^2}{(-x)^4} + {}_6C_5\frac{(2x)}{(-x)^5} + {}_6C_6\frac{1}{(-x)^6}$$
▲ +
$$= 64x^6 - 192x^4 + 240x^2 - 160 + 60x^{-2} - 12x^{-4} + x^{-6}$$
▲■   ▲+

The third term T_3 has the maximum coefficient 240 in the binomial expansion.

2.2 Finding the Sum of Coefficients

As a binomial expansion is an identity we can calculate the sum of the coefficients of the binomial expansion by setting two arguments of a binomial $(x+y)^n$ to appreciate values (... −2, −1, 0, 1, 2, ...). The sum of the first n terms of a sequence can be obtained by calculating the sum of all coefficients of the binomial expansion of a binomial $(x+y)^n$.

Example 2.2.1 The Sum of Coefficients

1) If $(x+3)^4 = a_4 x^4 + a_3 x^3 + a_2 x^2 + a_1 x + a_0$, $a_1 + a_2 + a_3 + a_4 = ?$
2) If $(2-x)^3 (1+2x)^2 = a_5 x^5 + a_4 x^4 + a_3 x^3 + a_2 x^2 + a_1 x + a_0$, $a_1 + a_2 + a_3 + a_4 + a_5 = ?$ and $a_0 + a_2 + a_4 = ?$
3) If $(4-x) + (4-x)^2 + (4-x)^3 + \cdots + (4-x)^n = a_0 + a_1 x + a_2 x^2 + a_3 x^3 + \cdots + a_n x^n$, $a_1 + a_2 + \cdots + a_n = ?$
4) If $(2\sqrt[3]{x} + \dfrac{1}{x^2})^n = a_0 + a_1 x + a_2 x^2 + \cdots + a_n x^n$ $(x \neq 0)$ and the term with x^2 is the fifth term, $a_1 + a_2 + \cdots + a_n = ?$

Solution:

1) Let $x = 0$, $a_0 = 3^4 = 81$.
   Let $x = 1$ $(1+3)^4 = a_4 + a_3 + a_2 + a_1 + a_0$.
   Then we have $a_1 + a_2 + a_3 + a_4 = 175$.

2) Let $x = 0$, $a_0 = 2^3 = 8$.
   Let $x = 1$, $3^2 = a_5 + a_4 + a_3 + a_2 + a_1 + a_0$.   (a)
   Then we have $a_1 + a_2 + a_3 + a_4 + a_5 = 1$.
   Let $x = -1$, $3^3 = -a_5 + a_4 - a_3 + a_2 - a_1 + a_0$.   (b)
   (a) + (b) $a_0 + a_2 + a_4 = 18$.

3) Let $x = 0$. $4 + 4^2 + 4^3 + \cdots + 4^n = a_0$
   then $a_0 = \dfrac{4(1-4^n)}{1-4}$.
   Let $x = 1$. $3 + 3^2 + 3^3 + \cdots + 3^n = a_0 + a_1 + a_2 + \cdots + a_n$.
   Then $a_0 + a_1 + a_2 + \cdots + a_n = \dfrac{3(1-3^n)}{1-3}$.
   $a_1 + a_2 + \cdots + a_n = \dfrac{3(1-3^n)}{1-3} - a_0$

$$= \frac{3(1-3^n)}{1-3} - \frac{4(1-4^n)}{1-4}$$
$$= \frac{3^{n+2}-2\cdot 4^{n+1}-1}{6}.$$

4) $T_{k+1} = {}_nC_k(2\sqrt[3]{x})^{n-k}(\frac{1}{x^2})^k = {}_nC_k \cdot 2^{n-k} x^{\frac{1}{3}(n-7k)}.$

Let $\frac{1}{3}(n-7k) = 2.$

Since $k = 4$ and $n = 34.$ $(2\sqrt[3]{x} + \frac{1}{x^2})^{34} = a_0 + a_1 x + a_2 x^2 + \cdots + a_{34} x^{34}.$

Let $x = 1,$ $a_0 + a_1 + a_2 + \cdots + a_{34} = 3^{34}.$

Example 2.2.2 The Sum of Coefficients

Prove the properties of the binomial coefficients.
1) The sum of all binomial coefficients of a binomial expansion is 2^n
$${}_nC_0 + {}_nC_1 + \cdots + {}_nC_k + \cdots + {}_nC_{n-1} + {}_nC_n = 2^n \qquad (n \in N^+)$$
2) The sum of binomial coefficients of all terms with odd index is equal to the sum of binomial coefficients of all terms with even index.
$${}_nC_1 + {}_nC_3 + {}_nC_5 + \cdots + {}_nC_{2n-1} + \cdots = {}_nC_0 + {}_nC_2 + {}_nC_4 + \cdots + {}_nC_{2n} + \cdots \quad (n \in N^+)$$
3) Either of two sums in question 2) is $2^{n-1}.$

Solution:
1) Suppose we have a binomial as below
$$(x+y)^n = {}_nC_0 x^n + {}_nC_1 x^{n-1} y + \cdots + {}_nC_i x^{n-i} y^i + \cdots + {}_nC_{n-1} x y^{n-1} + {}_nC_n y^n$$

Let $x = y = 1,$ then the binomial expansion becomes
$$2^n = {}_nC_0 + {}_nC_1 + \cdots + {}_nC_i + \cdots + {}_nC_{n-1} + {}_nC_n. \qquad (a)$$

2) The sum of binomial coefficients of all terms with odd index is
$${}_nC_1 + {}_nC_3 + {}_nC_5 + \cdots + {}_nC_{2n-1} + \cdots \qquad (b)$$
and the sum of binomial coefficients of all terms with even index is
$${}_nC_0 + {}_nC_2 + {}_nC_4 + \cdots + {}_nC_{2n} + \cdots. \qquad (c)$$

Let (b) − (c).
$$({}_nC_1 + {}_nC_3 + {}_nC_5 + \cdots + {}_nC_{2n-1} + \cdots) - ({}_nC_0 + {}_nC_2 + {}_nC_4 + \cdots + {}_nC_{2n} + \cdots)$$

2.2 Finding the Sum of Coefficients

We have
$$_nC_0 - {}_nC_1 + {}_nC_2 - {}_nC_3 + {}_nC_4 - {}_nC_5 + \cdots - {}_nC_{2n-1} + {}_nC_{2n} - \cdots. \quad (d)$$
We expect the expression (d) = 0.

Let's look at the binomial expansion of the binomial $(x+y)^n$.
$$(x+y)^n = {}_nC_0 x^n + {}_nC_1 x^{n-1} y + {}_nC_2 x^{n-2} y^2 + {}_nC_3 x^{n-3} y^3 + \cdots \quad (e)$$
Let $x = 1$ and $y = -1$, then (e) becomes
$$_nC_0 - {}_nC_1 + {}_nC_2 - {}_nC_3 + {}_nC_4 - {}_nC_5 + \cdots - {}_nC_{2n-1} + {}_nC_{2n} - \cdots = 0. \quad (f)$$
After comparing the formula (d) and (f) we obtain (d) = (f) = 0 then
$$_nC_1 + {}_nC_3 + {}_nC_5 + \cdots + {}_nC_{2n-1} + \cdots = {}_nC_0 + {}_nC_2 + {}_nC_4 + \cdots + {}_nC_{2n} + \cdots.$$

3) From the formula (a)
$$_nC_0 + {}_nC_1 + \cdots + {}_nC_i + \cdots + {}_nC_{n-1} + {}_nC_n = 2^n,$$
we have
$$({}_nC_1 + {}_nC_3 + {}_nC_5 + \cdots + {}_nC_{2n-1} + \cdots) + ({}_nC_0 + {}_nC_2 + {}_nC_4 + \cdots + {}_nC_{2n} + \cdots) = 2^n.$$
Since $\quad {}_nC_1 + {}_nC_3 + {}_nC_5 + \cdots + {}_nC_{2n-1} + \cdots = {}_nC_0 + {}_nC_2 + {}_nC_4 + \cdots + {}_nC_{2n} + \cdots,$
$$2({}_nC_1 + {}_nC_3 + {}_nC_5 + \cdots + {}_nC_{2n-1} + \cdots) = 2^n$$
or $\quad 2({}_nC_0 + {}_nC_2 + {}_nC_4 + \cdots + {}_nC_{2n} + \cdots) = 2^n.$

Then $\quad {}_nC_1 + {}_nC_3 + {}_nC_5 + \cdots + {}_nC_{2n-1} + \cdots = {}_nC_0 + {}_nC_2 + {}_nC_4 + \cdots + {}_nC_{2n} + \cdots = 2^{n-1}.$

Example 2.2.3   The Sum of Coefficients

Prove $1^2 + 3^2 + 5^2 + \cdots + (2n-1)^2 = \dfrac{1}{3} \cdot n \cdot (4n^2 - 1) \quad (n \in N^+)$

Solution:

Since the nth term of the left side of the formula is
$$(2n-1)^2 = 4n^2 - 4n + 1 = 4 \cdot n \cdot (n-1) + 1 = 8 \cdot {}_nC_2 + 1$$
we have
$$1^2 + 3^2 + 5^2 + \cdots + (2n-1)^2 = 1 + (8 \cdot {}_2C_2 + 1) + (8 \cdot {}_3C_2 + 1) + \cdots + (8 \cdot {}_nC_2 + 1).$$
$$= 8 \cdot ({}_2C_2 + {}_3C_2 + \cdots + {}_nC_2) + n.$$
By the identity of combinations, we have
$$_2C_2 + {}_3C_2 + \cdots + {}_nC_2 = {}_{n+1}C_3.$$
Thus $\quad 1^2 + 3^2 + 5^2 + \cdots + (2n-1)^2 = 8 \cdot {}_{n+1}C_3 + n = \dfrac{1}{3} \cdot n \cdot (4n^2 - 1).$

Example 2.2.4 The Sum of Coefficients

Find $1^3+2^3+3^3+\cdots+n^3 = ?$ $(n\in N^+)$

Solution:

Since $n^3=(n+1)\cdot n\cdot (n-1)+n=\dfrac{3!\cdot (n+1)\cdot n\cdot (n-1)}{3!}+n=6\cdot {}_{n+1}C_3+n,$

$$1^3+2^3+3^3+\cdots+n^3=1+(6\cdot {}_3C_3+2)+(6\cdot {}_4C_3+3)+\cdots+(6\cdot {}_{n+1}C_3+n)$$
$$=6\cdot ({}_3C_3+{}_4C_3+\cdots+{}_{n+1}C_3)+(1+2+3+\cdots+n).$$

By the identity of combinations, we have ${}_3C_3+{}_4C_3+\cdots+{}_{n+1}C_3={}_{n+2}C_4.$
Also we know

$$1+2+3+\cdots+n=\dfrac{n(n+1)}{2}.$$

Then

$$1^3+2^3+3^3+\cdots+n^3=6\cdot {}_{n+2}C_4+\dfrac{n(n+1)}{2}$$
$$=\dfrac{6\cdot (n+2)(n+1)n(n-1)}{4!}+\dfrac{n(n+1)}{2}$$
$$=\dfrac{(n+2)(n+1)n(n-1)}{4}+\dfrac{n(n+1)}{2}$$
$$=\dfrac{n^2(n+1)^2}{4}.$$

Example 2.2.5 The Sum of Coefficients

Find the sum of all coefficients of the binomial expansion of a binomial $(3-\sqrt[3]{x})^6$ except for that of the term containing x.

Solution:

The general term of the binomial $(3-\sqrt[3]{x})^6$ is

$$T_{k+1}={}_6C_k\, 3^{6-k}(-1)^k x^{\frac{k}{3}} \qquad (0\leqslant k\leqslant 6).$$

To find the term of x, let $\dfrac{k}{3}=1$ and we have $k = 3$. The coefficient of the term becomes

$$t_3=-{}_6C_3\, 3^3=-540.$$

To find the sum of all coefficients of the binomial expansion, let $x = 1$.

$$(3-\sqrt[3]{1})^6=64$$

Then the sum we are looking for becomes

$$64 - (-540) = 604.$$

2.2 Finding the Sum of Coefficients

Example 2.2.6   *The Sum of Coefficients*

Find the sum of all coefficients of terms with odd index in the expansion of the expression

$$(1+x)+(1+x)^2+(1+x)^3+\cdots+(1+x)^n.$$

Solution:

The expression consists of n binomials. By the property of binomial coefficients, the sum of coefficient of all terms with odd index in the binomial expansion of a binomial $(1+x)^n$ is 2^{n-1}.

Let S be the sum wanted and we have

$$S=1+2+2^2+2^3+\cdots+2^{n-2}+2^{n-1}.$$

As the formula is a geometric sequence we have

$$S=\frac{1\cdot(1-2^n)}{1-2}=2^n-1.$$

Example 2.2.7   *The Sum of Coefficients*

$2\cdot {}_{2n}C_0+{}_{2n}C_1+2\cdot {}_{2n}C_2+{}_{2n}C_3+\cdots+{}_{2n}C_{2n-1}+2\cdot {}_{2n}C_{2n}=?$

Solution:

$2\cdot {}_{2n}C_0+{}_{2n}C_1+2\cdot {}_{2n}C_2+{}_{2n}C_3+\cdots+{}_{2n}C_{2n-1}+2\cdot {}_{2n}C_{2n}$
$=({}_{2n}C_1+{}_{2n}C_3++\cdots+{}_{2n}C_{2n-3}+{}_{2n}C_{2n-1})+2({}_{2n}C_0+{}_{2n}C_2++\cdots+{}_{2n}C_{2n-2}+{}_{2n}C_{2n}).$

By the property of binomial coefficients, the above becomes

$$2^{2n-1}+2\cdot 2^{2n-1}.$$

Thus

$$2\cdot {}_{2n}C_0+{}_{2n}C_1+2\cdot {}_{2n}C_2+{}_{2n}C_3+\cdots+{}_{2n}C_{2n-1}+2\cdot {}_{2n}C_{2n}=3\cdot 2^{2n-1}.$$

Example 2.2.8   *The Sum of Coefficients*

If $4^n-{}_nC_1\cdot 4^{n-1}+{}_nC_2\cdot 4^{n-2}-\cdots+(-1)^{n-1}\cdot {}_nC_{n-1}\cdot 4+(-1)^n=729$, find $n=?$

Solution:

We will construct a binomial from the above formula.

$4^n-{}_nC_1\cdot 4^{n-1}+{}_nC_2\cdot 4^{n-2}-\cdots+(-1)^{n-1}\cdot {}_nC_{n-1}\cdot 4+(-1)^n$
$={}_nC_0 4^n+{}_nC_1\cdot 4^{n-1}\cdot(-1)+{}_nC_2\cdot 4^{n-2}\cdot(-1)^2-\cdots+{}_nC_{n-1}\cdot 4\cdot(-1)^{n-1}+{}_nC_n\cdot(-1)^n.$
$=(4-1)^n$
$=3^n.$

Let $3^n=729$ and we have $n=6$.

2.3 Proving Identities of Combinations

We have the following facts about a binomial expansion.
- a binomial expansion is an identity and
- the binomial coefficients of a binomial expansion are the numbers of combinations.

There are two ways to prove identities of combinations.
1. To assign a particular value to a binomial
   The binomial expansion of a binomial is
   $$(x+y)^n = {}_nC_0 x^n + {}_nC_1 x^{n-1} y + \cdots + {}_nC_k x^{n-k} y^k + \cdots + {}_nC_n y^n.$$
   If the terms x and y of the binomial are assigned by particular values respectively, we will get a series of identities of combinations.

2. To construct a new binomial
   Construct a binomial whose binomial expansion has appropriate form to match one side of the identity to be proven.

Example 2.3.1 Identities of Combinations

Prove the following identities of combinations.
1) ${}_nC_0 + {}_nC_1 + {}_nC_2 + \cdots + {}_nC_{n-1} + {}_nC_n = 2^n$
2) ${}_nC_0 - {}_nC_1 + {}_nC_2 + \cdots + (-1)^{n-1} {}_nC_{n-1} + (-1)^n {}_nC_n = 0$

Solution:
We use the binomial expansion of a binomial
$$(x+y)^n = {}_nC_0 x^n + {}_nC_1 x^{n-1} y + \cdots + \cdots + {}_nC_n y^n,$$
and assign proper values to the arguments x and y to prove identities of combinations.

1) Let $x = 1, y = 1$.
$$2^n = (1+1)^n = {}_nC_0 + {}_nC_1 + \cdots + {}_nC_{n-1} + {}_nC_n.$$

2) Let $x = 1, y = -1$.
$$0 = (1-1)^n = {}_nC_0 + (-1){}_nC_1 + \cdots + (-1)^{n-1} {}_nC_{n-1} + (-1)^n {}_nC_n = 0$$

If n is odd number,
$${}_nC_0 + {}_nC_2 + {}_nC_4 + \cdots + {}_nC_{n-1} = {}_nC_1 + {}_nC_3 + {}_nC_5 + \cdots + {}_nC_n.$$

2.3 Proving Identities of Combinations

If n is even number,

$$_nC_0 + {}_nC_2 + {}_nC_4 + \cdots + {}_nC_n = {}_nC_1 + {}_nC_3 + {}_nC_5 + \cdots + {}_nC_{n-1}.$$

Then the following identity holds.

$$_nC_1 + {}_nC_3 + {}_nC_5 + \cdots = {}_nC_0 + {}_nC_2 + {}_nC_4 + \cdots = 2^{n-1}.$$

Example 2.3.2 Identities of Combinations

Prove
1) $2\,{}_nC_1 + 2^2\,{}_nC_2 + \cdots + 2^n\,{}_nC_n = 3^n - 1$
2) $3^n\,{}_nC_0 + 3^{n-1}\,{}_nC_1 + 3^{n-2}\,{}_nC_2 + \cdots + {}_nC_n = 4^n$

Solution:

1) The left side of the formula can be written to

$$_nC_0 + 2\,{}_nC_1 + 2^2\,{}_nC_2 + \cdots + 2^n\,{}_nC_n - 1.$$

Since
$$(1+2)^n = {}_nC_0 + 2\,{}_nC_1 + 2^2\,{}_nC_2 + \cdots + 2^n\,{}_nC_n,$$
$$_nC_0 + 2\,{}_nC_1 + 2^2\,{}_nC_2 + \cdots + 2^n\,{}_nC_n - 1 = 3^n - 1$$

Thus
$$2\,{}_nC_1 + 2^2\,{}_nC_2 + \cdots + 2^n\,{}_nC_n = 3^n - 1.$$

2) Let $x=3$, $y=1$,
then
$$(3+1)^n = {}_nC_0\,3^n + {}_nC_1\,3^{n-1} + \cdots + {}_nC_k\,3^{n-k} + \cdots + {}_nC_n$$
$$3^n\,{}_nC_0 + 3^{n-1}\,{}_nC_1 + 3^{n-2}\,{}_nC_2 + \cdots + {}_nC_n = 4^n.$$

Example 2.3.3 Identities of Combinations

Prove $_nC_1 + 2\cdot{}_nC_2 + 3\cdot{}_nC_3 + \cdots + n\cdot{}_nC_n = n\cdot 2^{n-1}.$

Solution:
Using the identity of combinations,

$$_nC_m = \frac{n}{m}\cdot{}_{n-1}C_{m-1},$$

we arrange the left side of the identity as below.

$$n\cdot{}_{n-1}C_0 + n\cdot{}_{n-1}C_1 + n\cdot{}_{n-1}C_2 + \cdots + n\cdot{}_{n-1}C_{n-1}$$
$$= n\cdot({}_nC_0 + n\cdot{}_{n-1}C_1 + n\cdot{}_{n-1}C_2 + \cdots + n\cdot{}_{n-1}C_{n-1})$$

As
$$_nC_0 + n\cdot{}_{n-1}C_1 + n\cdot{}_{n-1}C_2 + \cdots + n\cdot{}_{n-1}C_{n-1} = 2^{n-1},$$
$$_nC_1 + 2\cdot{}_nC_2 + 3\cdot{}_nC_3 + \cdots + n\cdot{}_nC_n = n\cdot 2^{n-1}.$$

Example 2.3.4　Identities of Combinations

Prove $\dfrac{_nC_0}{1}+\dfrac{_nC_1}{2}+\dfrac{_nC_3}{3}+\cdots+\dfrac{_nC_n}{n+1}=\dfrac{1}{n+1}(2^{n+1}-1)$.

Solution:

By the identity of combinations,

$$_nC_m=\dfrac{m+1}{n+1}\cdot {_{n+1}C_{m+1}},$$

we arrange the left side of the above as below.

$$\dfrac{_nC_0}{1}+\dfrac{_nC_1}{2}+\dfrac{_nC_3}{3}+\cdots+\dfrac{_nC_n}{n+1}$$

$$=\dfrac{1}{n+1}{_{n+1}C_1}+\dfrac{1}{n+1}{_{n+1}C_2}+\dfrac{1}{n+1}{_{n+1}C_3}+\cdots+\dfrac{1}{n+1}{_{n+1}C_{n+1}}$$

$$=\dfrac{1}{n+1}({_{n+1}C_1}+{_{n+1}C_2}+{_{n+1}C_3}+\cdots+{_{n+1}C_{n+1}})$$

$$=\dfrac{1}{n+1}({_{n+1}C_0}+{_{n+1}C_1}+{_{n+1}C_2}+{_{n+1}C_3}+\cdots+{_{n+1}C_{n+1}}-1)$$

By property of binomial coefficients,

$$_{n+1}C_0+{_{n+1}C_1}+{_{n+1}C_2}+{_{n+1}C_3}+\cdots+{_{n+1}C_{n+1}}=2^{n+1},$$

then

$$\dfrac{_nC_0}{1}+\dfrac{_nC_1}{2}+\dfrac{_nC_3}{3}+\cdots+\dfrac{_nC_n}{n+1}=\dfrac{1}{n+1}(2^{n+1}-1).$$

Example 2.3.5　Identities of Combinations

Prove $1+4\,_nC_1+7\,_nC_2+10\,_nC_3+\cdots+(3n+1)\,_nC_n=(3n+2)\cdot 2^{n-1}$

Solution:

Let $\quad S=1+4\,_nC_1+7\,_nC_2+10\,_nC_3+\cdots+(3n-2)\,_nC_{n-1}+(3n+1)\,_nC_n\quad$ (a)

Reverse (a)

$$S=(3n+1)\,_nC_n+(3n-2)+\cdots+10\,_nC_3+7\,_nC_2+4\,_nC_1+1 \qquad\text{(b)}$$

Add (a) and (b)

$$2\cdot S=(1+(3n+1)\,_nC_n)+(4\,_nC_1+(3n-2)\,_nC_{n-1})+\cdots. \qquad\text{(c)}$$

As $1={_nC_n},\ {_nC_1}={_nC_{n-1}},\ {_nC_2}={_nC_{n-2}},\ \cdots,$

$$2\cdot S=(3n+2)\cdot({_nC_n}+{_nC_{n-1}}+{_nC_{n-2}}+\cdots+{_nC_2}+{_nC_1}+{_nC_0}).$$

By the property of binomial coefficients,

$$_nC_n+{_nC_{n-1}}+{_nC_{n-2}}+\cdots+{_nC_2}+{_nC_1}+{_nC_0}=2^n.$$

We have $\qquad 2\cdot S=(3n+2)\cdot 2^n.$

Thus $\qquad 1+4\,_nC_1+7\,_nC_2+10\,_nC_3+\cdots+(3n+1)\,_nC_n=(3n+2)\cdot 2^{n-1}.$

2.3 Proving Identities of Combinations

Example 2.3.6   Identities of Combinations

Prove $2\,_nC_0 + \dfrac{2^2}{2}\cdot{}_nC_1 + \dfrac{2^3}{3}\cdot{}_nC_2 + \cdots + \dfrac{2^{n+1}}{n+1}\cdot{}_nC_n = \dfrac{3^{n+1}-1}{n+1}.$

Solution:

Construct a binomial like $(x+y)^{n+1}$ and

$$(x+y)^{n+1} = {}_{n+1}C_0\, x^{n+1} + {}_{n+1}C_1\, x^n y + \cdots + {}_{n+1}C_n\, x\, y^n + {}_{n+1}C_{n+1}\, y^{n+1}.$$

Let $x = 1$ and $y = 2$. We have

$$1 + 2\,_{n+1}C_1 + 2^2\,_{n+1}C_2 + \cdots + 2^n\,_{n+1}C_n + 2^{n+1}\,_{n+1}C_{n+1} = 3^{n+1}$$

$$2\,_{n+1}C_1 + 2^2\,_{n+1}C_2 + \cdots + 2^n\,_{n+1}C_n + 2^{n+1}\,_{n+1}C_{n^2+1} = 3^{n+1} - 1.$$

Divide both sides by $n + 1$.

$$\dfrac{2}{n+1}\,_{n+1}C_1 + \dfrac{2^2}{n+1}\,_{n+1}C_2 + \cdots + \dfrac{2^n}{n+1}\,_{n+1}C_n + \dfrac{2^{n+1}}{n+1}\,_{n+1}C_{n+1} = \dfrac{3^{n+1}-1}{n+1} \quad \text{(a)}$$

Since

$$\dfrac{2^k}{n+1}\,_{n+1}C_k = \dfrac{2^k}{k}\,_nC_{k-1} \quad (k = 1, 2, \ldots, n+1),$$

the left side of the formula (a) above is

$$\dfrac{2\,_{n+1}C_1}{n+1} + \dfrac{2^2\,_{n+1}C_2}{n+1} + \cdots + 2^n\dfrac{_{n+1}C_n}{n+1} + 2^{n+1}\dfrac{_{n+1}C_{n+1}}{n+1}$$

$$= 2\,_nC_0 + \dfrac{2^2}{2}\,_nC_1 + \cdots + \dfrac{2^n}{n}\,_nC_{n-1} + \dfrac{2^{n+1}}{n+1}\,_nC_n.$$

Then we obtain

$$2\,_nC_0 + \dfrac{2^2}{2}\,_nC_1 + \dfrac{2^3}{3}\,_nC_2 + \cdots + \dfrac{2^{n+1}}{n+1}\,_nC_n = \dfrac{3^{n+1}-1}{n+1}.$$

2.4 Proving Inequalities

Below is a guideline to prove inequalities using binomial expansion.
1) Arrange one side of a inequality to a proper binomial and scale the inequality if necessary.
2) Expand the binomial.
3) Research the binomial expansion and compare it with the other side of the inequality.

Example 2.4.1 *Proving Inequality*
Prove the following inequalities ($n \in N^+$, $n \geqslant 2$).
1) $4^n > 3^{n-1} \cdot (3+n)$ 2) $\dfrac{n(n-1)}{2} < 2^n$ 3) $3^n > 2^{n-1} \cdot (n+2)$ 4) $2 < \left(1+\dfrac{1}{n}\right)^n$

Solution:
1) From the left side of the inequality we have
$$4^n = (3+1)^n = {}_nC_0 \cdot 3^n + {}_nC_1 \cdot 3^{n-1} + {}_nC_2 \cdot 3^{n-2} + \cdots + {}_nC_n$$
$$= 3^n + n \cdot 3^{n-1} + ({}_nC_2 \cdot 3^{n-2} + \cdots + {}_nC_n)$$
$$= 3^{n-1} \cdot (3+n) + ({}_nC_2 \cdot 3^{n-2} + \cdots + {}_nC_n).$$

Because
$${}_nC_2 \cdot 3^{n-2} + \cdots + {}_nC_n > 0,$$
$$4^n > 3^{n-1} \cdot (3+n).$$

2) By the property of binomial coefficients, we known
$${}_nC_0 + {}_nC_1 + {}_nC_2 + \cdots + {}_nC_{n-1} + {}_nC_n = 2^n.$$
All the terms of the left side of above identity are positive. Since $n \geqslant 2$, there are at least $n+1=3$ terms in the binomial expansion.

As the third term
$${}_nC_2 = \dfrac{n(n-1)}{2},$$
$$\dfrac{n(n-1)}{2} < 2^n.$$

3) From the left side of the inequality we have
$$3^n = (2+1)^n = {}_nC_0 \cdot 2^n + {}_nC_1 \cdot 2^{n-1} + {}_nC_2 \cdot 2^{n-2} + \cdots + {}_nC_{n-1} 2 + {}_nC_n$$
$$= 2^n + n \cdot 2^{n-1} + ({}_nC_2 \cdot 2^{n-2} + \cdots + {}_nC_{n-1} 2 + {}_nC_n)$$
$$= 2^{n-1}(n+2) + ({}_nC_2 \cdot 2^{n-2} + \cdots + {}_nC_{n-1} 2 + {}_nC_n).$$

2.4 Proving Inequalities

Because
$$_nC_2 \cdot 2^{n-2} + \cdots + {_nC_{n-1}} 2 + {_nC_n} > 0,$$
$$3^n > 2^{n-1} \cdot (n+2).$$

4) Expand the binomial from the right side of the expression.
$$\left(1+\frac{1}{n}\right)^n = {_nC_0} + {_nC_1} \cdot \frac{1}{n} + {_nC_2} \cdot \left(\frac{1}{n}\right)^2 + \cdots + {_nC_n} \cdot \left(\frac{1}{n}\right)^n$$
$$= 2 + {_nC_2} \cdot \left(\frac{1}{n}\right)^2 + \cdots + {_nC_n} \cdot \left(\frac{1}{n}\right)^n$$

Since all the terms on the right side are positive when $n \geq 2$, the expansion has at least $n+1=3$ terms.

Because
$$_nC_2 \cdot \left(\frac{1}{n}\right)^2 + \cdots + {_nC_n} \cdot \left(\frac{1}{n}\right)^n > 0,$$
$$2 + {_nC_2} \cdot \left(\frac{1}{n}\right)^2 + \cdots + {_nC_n} \cdot \left(\frac{1}{n}\right)^n > 2.$$

Therefore
$$2 < \left(1+\frac{1}{n}\right)^n.$$

Example 2.4.2 *Proving Inequality*

Prove $(n+1)^n \geq 2n^n + (n-1)^n$ $(n \in N^*, n \geq 2)$.

Solution:

Arrange the inequality to
$$(n+1)^n - (n-1)^n \geq 2n^n.$$

The left side of the inequality above becomes
$$n^n\left(\left(1+\frac{1}{n}\right)^n - \left(1-\frac{1}{n}\right)^n\right)$$
$$= n^n\left(\left({_nC_0} + {_nC_1}\frac{1}{n} + {_nC_2}\frac{1}{n^2} + {_nC_3}\frac{1}{n^3} + \cdots\right) - \left({_nC_0} - {_nC_1}\frac{1}{n} + {_nC_2}\frac{1}{n^2} - {_nC_3}\frac{1}{n^3} + \cdots\right)\right)$$
$$= 2 \cdot n^n \underbrace{\left(1 + {_nC_3}\frac{1}{n^3} + {_nC_5}\frac{1}{n^5} + {_nC_7}\frac{1}{n^7} + \cdots\right)}_{>1}.$$

Because
$$\left(1 + {_nC_3}\frac{1}{n^3} + {_nC_5}\frac{1}{n^5} + {_nC_7}\frac{1}{n^7} + \cdots\right) > 1,$$
$$(n+1)^n \geq 2n^n + (n-1)^n.$$

Example 2.4.3   Proving Inequality

If $x > 0$, $y > 0$, and $x \neq y$, prove that
$$\frac{x^n + y^n}{2} > \left(\frac{x+y}{2}\right)^n \quad (n > 1).$$

Solution:

Let
$$x = \frac{x+y}{2} + \frac{x-y}{2}$$

and
$$y = \frac{x+y}{2} - \frac{x-y}{2}.$$

The left side of the inequality becomes

$$\frac{x^n + y^n}{2} = \frac{1}{2}\left[\left(\frac{x+y}{2} + \frac{x-y}{2}\right)^n + \left(\frac{x+y}{2} - \frac{x-y}{2}\right)^n\right]$$

$$= \frac{1}{2}\left({}_nC_0\left(\frac{x+y}{2}\right)^n + {}_nC_1\left(\frac{x+y}{2}\right)^{n-1} + {}_nC_2\left(\frac{x+y}{2}\right)^{n-2} + {}_nC_3\left(\frac{x+y}{2}\right)^{n-3} + \cdots\right)$$

$$+ \frac{1}{2}\left({}_nC_0\left(\frac{x+y}{2}\right)^n - {}_nC_1\left(\frac{x+y}{2}\right)^{n-1} + {}_nC_2\left(\frac{x+y}{2}\right)^{n-2} - {}_nC_3\left(\frac{x+y}{2}\right)^{n-3} + \cdots\right)$$

$$= \left(\frac{x+y}{2}\right)^n + \underbrace{{}_nC_2\left(\frac{x+y}{2}\right)^{n-2} + {}_nC_4\left(\frac{x+y}{2}\right)^{n-4} + {}_nC_6\left(\frac{x+y}{2}\right)^{n-6} + \cdots}_{> 0}$$

Because ${}_nC_2\left(\frac{x+y}{2}\right)^{n-2} + {}_nC_4\left(\frac{x+y}{2}\right)^{n-4} + {}_nC_6\left(\frac{x+y}{2}\right)^{n-6} + \cdots > 0,$

$$\frac{x^n + y^n}{2} > \left(\frac{x+y}{2}\right)^n.$$

2.5 Factoring Algebraic Expressions

Binomial expansion of a binomial can be used to factor a algebraic expression.

> **Factoring Algebraic Expressions**
>
> A algebraic expression $g(x)$ is a factor of another algebraic expression $f(x)$ if $g(x)$ is a factor of each term of the binomial expansion of $f(x)$.

Below is the procedure to find the factor $g(x)$ and the remainder of a algebraic expression $f(x)$.

1) Arrange $f(x)$ to a binomial whose one of two terms is a multiple of $g(x)$.
2) Expand the binomial.
3) Separate the binomial expansion into two parts, one is an integer which is a multiple of $g(x)$ and the other is a remainder.
4) Recursively repeat steps above on the remainder if it is greater than $g(x)$.
5) If no remainder left $g(x)$ is a factor of $f(x)$ else the last remainder part is the remainder of $f(x)$.

Example 2.5.1 Factoring Algebraic Expressions

1) Prove that n^2 is a factor of the expression $(n+1)^n - 1$, $n \in N^+$.
2) If $(x-1)$ is a factor of the expression $x^6 + a$ find $a = ?$

Solution:

1) Because $(n+1)^n$ is a binomial

$$(n+1)^n - 1 = {}_nC_0 n^n + {}_nC_1 n^{n-1} + \cdots + {}_nC_{n-2} n^2 + {}_nC_{n-1} n + {}_nC_n - 1.$$

As ${}_nC_n = 1$ and ${}_nC_{n-1} = n$, we have

$$(n+1)^n - 1 = {}_nC_0 n^n + {}_nC_1 n^{n-1} + \cdots + {}_nC_{n-2} n^2 + n^2$$

$$= \underbrace{n^2 ({}_nC_0 n^{n-2} + {}_nC_1 n^{n-3} + \cdots + {}_nC_{n-2} + 1)}_{\text{multiple of } n^2}.$$

No remainder is left, then n^2 is a factor of the expression $(n+1)^n - 1$.

2) Arrange $\qquad x^6 + a = ((x-1) + 1)^6 + a$

to let one of arguments of $((x-1)+1)^6$ be a multiple of $(x-1)$.

$$((x-1)+1)^6 + a$$
$$= {}_6C_0(x-1)^6 + {}_6C_1(x-1)^5 + {}_6C_2(x-1)^4 + {}_6C_3(x-1)^3 + {}_6C_4(x-1)^2 + {}_6C_5(x-1) + {}_6C_6 + a$$
$$= \underbrace{(x-1)\cdot[(x-1)^5 + 6(x-1)^4 + 15(x-1)^3 + 20(x-1)^2 + 15(x-1) + 6]}_{\text{multiple of }(x-1)} + \underbrace{1+a}_{\text{remainder}}$$

Since $(x-1)$ is a factor of $x^6 + a$, the remainder does not exist. Let $1+a=0$ and we obtain $a=-1$.

Example 2.5.2 Factoring Algebraic Expressions
Prove that 5 is a factor of the expression $31^{13} - 1$.

Solution:
At first we construct a binomial whose one term is a multiple of 5, say 30.
$$31^{13} - 1 = (30+1)^{13} - 1$$
$$= ({}_{13}C_0 \cdot 30^{13} + {}_{13}C_1 \cdot 30^{12} + {}_{13}C_2 \cdot 30^{11} + \cdots + {}_{13}C_{12} \cdot 30 + {}_{13}C_{13}) - 1.$$
$$= 30({}_{13}C_0 \cdot 30^{12} + {}_{13}C_1 \cdot 30^{11} + {}_{13}C_2 \cdot 30^{10} + \cdots + {}_{13}C_{12})$$

Because 30 is a multiple of 5 and
$$({}_{13}C_0 \cdot 30^{12} + {}_{13}C_1 \cdot 30^{11} + {}_{13}C_2 \cdot 30^{10} + \cdots + {}_{13}C_{12})$$
is an integer, 5 is a factor of $31^{13} - 1$.

Example 2.5.3 Factoring Algebraic Expressions
Prove that 36 is a factor of expression $7^n - 6n - 1$.

Solution:
Arrange
$$7^n - 6n - 1 = (1+6)^n - 6n - 1 \quad \text{(a)}$$

Since the binomial expansion of the binomial $(1+6)^n$ is
$$(1+6)^n = {}_nC_0 + {}_nC_1 \cdot 6 + {}_nC_2 \cdot 6^2 + {}_nC_3 \cdot 6^3 + \cdots + {}_nC_{n-1} \cdot 6^{n-1} + {}_nC_n \cdot 6^n$$

the expression (a) becomes
$$7^n - 6n - 1 = ({}_nC_0 + {}_nC_1 \cdot 6 + {}_nC_2 \cdot 6^2 + {}_nC_3 \cdot 6^3 + \cdots + {}_nC_{n-1} \cdot 6^{n-1} + {}_nC_n \cdot 6^n) - 6n - 1.$$
$$= {}_nC_2 \cdot 6^2 + {}_nC_3 \cdot 6^3 + {}_nC_4 \cdot 6^4 + \cdots + {}_nC_{n-1} \cdot 6^{n-1} + {}_nC_n \cdot 6^n$$
$$= 36\underbrace{({}_nC_2 + {}_nC_3 \cdot 6 + {}_nC_4 \cdot 6^2 + \cdots + {}_nC_{n-1} \cdot 6^{n-3} + {}_nC_n \cdot 6^{n-2})}_{\text{multiple of 36}}.$$

Because $({}_nC_2 + {}_nC_3 \cdot 6 + {}_nC_4 \cdot 6^2 + \cdots + {}_nC_{n-1} \cdot 6^{n-3} + {}_nC_n \cdot 6^{n-2})$ is an integer, 36 is a factor of expression $7^n - 6n - 1$.

2.5 Factoring Algebraic Expressions

Example 2.5.4   *Factoring Algebraic Expressions*

Find the remainder of the expression $\dfrac{13^4}{3}$.

Solution:

Arrange 13^4 to a binomial $(12+1)^4$ and expand the binomial

$$(12+1)^4 = {_4C_0}\,12^4 + {_4C_1}\,12^3 + {_4C_2}\,12^2 + {_4C_3}\,12 + {_4C_4}$$
$$= 12 \cdot (\,{_4C_0}\cdot 12^3 + {_4C_1}\cdot 12^2 + {_4C_2}\cdot 12 + {_4C_3}) + 1$$

$\underbrace{\phantom{12 \cdot ({_4C_0}\cdot 12^3 + {_4C_1}\cdot 12^2 + {_4C_2}\cdot 12 + {_4C_3})}}_{\text{multiple of 3}}$ $\underbrace{}_{\text{remainder} < 3}$

The first part $12 \cdot ({_4C_0}\cdot 12^3 + {_4C_1}\cdot 12^2 + {_4C_2}\cdot 12 + {_4C_3})$ is a multiple of 3 and the second part, the remainder, is *1* and smaller than *3*. So we don't need proceed further and the remainder of the expression $\dfrac{13^4}{3}$ is *1*.

Example 2.5.5   *Factoring Algebraic Expressions*

Find the remainder of the expression $\dfrac{27^5}{4}$.

Solution:

Arrange 27^5 to a binomial $(24+3)^5$ and

$$(24+3)^5 = {_5C_0}\,24^5 + {_5C_1}\,24^4 \cdot 3 + {_5C_2}\,24^3 \cdot 3^2 + {_5C_3}\,24^2 \cdot 3^3 + {_5C_4}\,24 \cdot 3^4 + {_5C_5}\cdot 3^5$$
$$= 24 \cdot ({_5C_0}\,24^4 + {_5C_1}\,24^3 \cdot 3 + {_5C_2}\,24^2 \cdot 3^2 + {_5C_3}\,24 \cdot 3^3 + {_5C_4}\cdot 3^4) + {_5C_5}\cdot 3^5$$

$\underbrace{\phantom{= 24 \cdot ({_5C_0}\,24^4 + {_5C_1}\,24^3 \cdot 3 + {_5C_2}\,24^2 \cdot 3^2 + {_5C_3}\,24 \cdot 3^3 + {_5C_4}\cdot 3^4)}}_{\text{multiple of 4}}$ $\underbrace{\phantom{{_5C_5}\cdot 3^5}}_{\text{remainder} > 4}$

The first part of the above,

$$24 \cdot ({_5C_0}\,24^4 + {_5C_1}\,24^3 \cdot 3 + {_5C_2}\,24^2 \cdot 3^2 + {_5C_3}\,24 \cdot 3^3 + {_5C_4}\cdot 3^4),$$

is a multiple of 4. The second part, the reminder, 3^5 is greater than 4, so we continue factoring on the remainder in the same way.

$$3^5 = (4-1)^5 = {_5C_0}\,4^5 - {_5C_1}\,4^4 + {_5C_2}\,4^3 - {_5C_3}\,4^2 + {_5C_3}\,4 - {_5C_5}$$
$$= 4({_5C_0}\,4^4 - {_5C_1}\,4^3 + {_5C_2}\,4^2 - {_5C_3}\,4 + {_5C_3}) - 1$$

$\underbrace{\phantom{= 4({_5C_0}\,4^4 - {_5C_1}\,4^3 + {_5C_2}\,4^2 - {_5C_3}\,4 + {_5C_3})}}_{\text{multiple of 4}}$

The last term is *–1*. As a remainder can not be negative we borrow one *4* from the integer part then the remainder of the expression $\dfrac{3^5}{4}$ should be $4 - 1 = 3$.

Therefore the remainder of the expression $\dfrac{27^{15}}{4}$ is *3*.

Example 2.5.6 **Factoring Algebraic Expressions**
If $(x-1)^2$ is a factor of the expression x^6-px+q, find p and q.
Solution:

$x^6 = (1+(x-1))^6 =$
$_6C_0 + {_6C_1}(x-1) + {_6C_2}(x-1)^2 + {_6C_3}(x-1)^3 + {_6C_2}(x-1)^4 + {_6C_5}(x-1)^5 + {_6C_6}(x-1)^6$
$= 1 + 6(x-1) + (x-1)^2({_6C_2} + {_6C_3}(x-1) + {_6C_2}(x-1)^2 + {_6C_5}(x-1)^3 + {_6C_6}(x-1)^4),$

$x^6 - px + q =$
$\underbrace{(x-1)^2({_6C_2} + {_6C_3}(x-1) + {_6C_2}(x-1)^2 + {_6C_5}(x-1)^3 + {_6C_6}(x-1)^4)}_{\text{multiple of }(x-1)^2} + \underbrace{(6-p)x + 5 + q}_{\text{remainder}}$

Because $(x-1)^2$ is a factor of the expression $x^6 - px + q$, the remainder should be zero.

Let $(6-p)x + 5 + q = 0.$

As this formula applies to any x, we obtain $q = -5$ and $p = 6$ by setting $x = 0$ and $x = 1$ respectively.

Example 2.5.7 **Factoring Algebraic Expressions**
If today is Monday, what day is the last day of 3^{100} days after today?
Solution:
Arrange
$3^{100} = (7+2)^{50}$
$= {_{50}C_0}7^{50} + {_{50}C_1}7^{49}\cdot 2 + \cdots + {_{50}C_{49}}7\cdot 2^{49} + {_{50}C_{50}}\cdot 2^{50}$
$= 7\cdot\underbrace{({_{50}C_0}7^{49} + {_{50}C_1}7^{48}\cdot 2 + \cdots + {_{50}C_{49}}2^{49})}_{\text{multiple of 7}} + \underbrace{2^{50}}_{\text{remainder}}$

Then we continue to factor the remainder.
$2^{50} = (2^3)^{18}\cdot 2^2 = (7+1)^{18}\cdot 2^2$
$= 7\cdot\underbrace{({_{18}C_0}7^{17} + {_{18}C_1}7^{16} + \cdots + {_{18}C_{17}})\cdot 2^2}_{\text{multiple of 7}} + \underbrace{{_{18}C_{18}}\cdot 2^2}_{\text{remainder}}$

The remainder equals to 4 and that day is Friday.

2.5 Factoring Algebraic Expressions

Example 2.5.8 *Factoring Algebraic Expressions*

How many consecutive zeros at the tail of the number $11^{10}-1$?

Solution:

$$11^{10}-1=(10-1)^{10}$$
$$={}_{10}C_0\cdot 10^{10}+{}_{10}C_1\cdot 10^9+\cdots+{}_{10}C_8\cdot 10^2+{}_{10}C_9\cdot 10+{}_{10}C_{10}-1$$
$$=10^{10}+{}_{10}C_1\cdot 10^9+\cdots+{}_{10}C_8\cdot 10^2+\underbrace{{}_{10}C_9\cdot 10}_{=100}$$

There are two consecutive zeros at the tail of the expression $11^{10}-1$.

Example 2.5.9 *Factoring Algebraic Expressions*

Prove that *64* is a factor of expression $3^{2n+2}-8n-9$.

Solution:

Arrange $3^{2n+2}-8n-9$
to $3^{2(n+1)}-8n-9$.

We have

$$3^{2(n+1)}-8n-9$$
$$=9\cdot(8+1)^n-8n-9$$
$$=9\cdot({}_nC_0 8^n+{}_nC_1 8^{n-1}+\cdots+{}_nC_{n-1}8+{}_nC_n)-8n-9$$
$$=9\cdot({}_nC_0 8^n+{}_nC_1 8^{n-1}+\cdots+{}_nC_{n-2}8^2)+9\cdot{}_nC_{n-1}8+9\cdot{}_nC_n-8n-9$$
$$=64\cdot 9\cdot({}_nC_0 8^{n-2}+{}_nC_1 8^{n-3}+\cdots+{}_nC_{n-2})+64\cdot n$$
$$=64\cdot[9\cdot({}_nC_0 8^{n-2}+{}_nC_1 8^{n-3}+\cdots+{}_nC_{n-2})+n].$$

Therefore *64* is a factor of expression $3^{2n+2}-8n-9$.

2.6 Finding General Term of a Polynomial Expansion

The following procedure shows you how to find the general term of a polynomial by the binomial theorem.
1. Arrange a polynomial to a binomial and find the general term of the binomial expansion.
2. If any binomial exists within the formula of the general term above find the general term for that binomial too.
3. A compound formula of the general term for the polynomial expansion is obtained.

For example, a polynomial $(3+2x-x^2)^n$ can be arranged to one of these three binomials

$$[3+(2x-x^2)]^n$$
$$[(3+2x)-x^2]^n$$
$$[2x+(3-x^2)]^n$$

To make your computation easy, we need to select a proper one.

Example 2.6.1 Find General Term of Polynomial Expansions

Find the coefficient of the term x^4 of polynomial expansion of the polynomial $(1+x+x^3)^6$.

Solution:

Arrange the polynomial $(1+x+x^3)^6$ to the binomial
$$[1+(x+x^3)]^6.$$

The general term of binomial expansion of the binomial is
$$T_{k+1} = {_6}C_k(x+x^3)^k = {_6}C_k x^k (1+x^2)^k \quad (0 \leqslant k \leqslant 6).$$

Let T'_{p+1} be the general term of binomial expansion of the binomial $(1+x^2)^k$ above.
$$T'_{p+1} = {_k}C_p x^{2p}.$$

Let $T(k, p)$ represent T_{k+1} then
$$T(k, p) = {_6}C_k \cdot {_k}C_p \cdot x^{k+2p}$$

2.6 Finding General Term of a Polynomial Expansion

Let $\quad k+2p = 4 \quad (0 \leqslant k \leqslant 6, 0 \leqslant p \leqslant k)$
To meet the condition above, we have two potential pairs of (k, p).
$$(2, 1), (4, 0)$$
There are two terms of x^4 in the polynomial expansion before combining like terms,
$$T(2,1) = {}_6C_2 \cdot {}_2C_1 \cdot x^4, \quad T(4,0) = {}_6C_4 \cdot {}_4C_0 \cdot x^4.$$
After combining like terms the term x^4 becomes
$$T(2,1) + T(4,0) = {}_6C_2 \cdot {}_2C_1 \cdot x^4 + {}_6C_4 \cdot {}_4C_0 \cdot x^4$$
$$= ({}_6C_2 \cdot {}_2C_1 + {}_6C_4 \cdot {}_4C_0) \cdot x^4.$$

The coefficient of the term x^4 is ${}_6C_2 \cdot {}_2C_1 + {}_6C_4 \cdot {}_4C_0 = 45$.

Example 2.6.2 Find General Term of Polynomial Expansions

Given the polynomial $(3a-b+2c+d)^8$, find
1) the term $a^2 \cdot b^2 \cdot c \cdot d^3$ in the polynomial expansion and
2) the coefficient of the term $a \cdot b^6 \cdot c$.

Solution:
1) Arrange the polynomial $(3a-b+2c+d)^8$ to the binomial
$$[(3a-b)+(2c+d)]^8.$$

The general term of the binomial expansion of $[(3a-b)+(2c+d)]^8$ is
$$T_{k+1} = {}_8C_k (3a-b)^{8-k} \cdot (2c+d)^k \qquad (0 \leqslant k \leqslant 8).$$

Let T'_{k+1} be the general term of the binomial expansion of the binomial $(3a-b)^{8-k}$.
$$T'_{p+1} = {}_{8-k}C_p (3a)^{8-k-p} \cdot (-b)^p \qquad (0 \leqslant p \leqslant 8-k)..$$

Let T'_{k+1} be the general term of the binomial expansion of the binomial $(2c+d)^k$.
$$T'_{q+1} = {}_kC_q (2c)^{k-q} \cdot d^q \qquad (0 \leqslant q \leqslant k).$$

Let $T(k, p, q)$ represent T_{k+1} then
$$T(k,p,q) = {}_8C_k T'_{p+1} \cdot T'_{q+1}$$
$$= {}_8C_k \cdot [{}_{8-k}C_p (3a)^{8-k-p} \cdot (-b)^p] \cdot [{}_kC_q (2c)^{k-q} \cdot d^q]$$

$$= (-1)^p \cdot {}_8C_k \cdot {}_{8-k}C_p \cdot {}_kC_q 3^{8-k-p} \cdot 2^{k-q} \cdot a^{8-k-p} \cdot b^p \cdot c^{k-q} \cdot d^q.$$

Let
$$\begin{cases} q = 3 \\ k - q = 1 \\ p = 2 \\ 8 - k - p = 2 \end{cases} \quad \begin{array}{l} (0 \leqslant k \leqslant 8,\ 0 \leqslant q \leqslant k) \\ \\ (0 \leqslant k \leqslant 8,\ 0 \leqslant p \leqslant 8-k) \end{array}$$

After solving the equation set above we have one solution of (k, p, q)
$$(4, 2, 3).$$

Thus the term $a^2 \cdot b^2 \cdot c \cdot d^3$ becomes
$$T(4, 2, 3) = (-1)^2 \cdot {}_8C_4 \cdot {}_{8-4}C_2 \cdot {}_4C_3 \, 3^{8-4-2} \cdot 2^{4-3} \cdot a^{8-4-2} \cdot b^2 \cdot c^{4-3} \cdot d^3.$$
$$= 18 \cdot {}_8C_4 \cdot {}_4C_2 \cdot {}_4C_3 \cdot a^2 \cdot b^2 \cdot c \cdot d^3$$
$$= 30{,}240 \cdot a^2 \cdot b^2 \cdot c \cdot d^3.$$

2) Let
$$\begin{cases} q = 0 \\ k - q = 1 \\ p = 6 \\ 8 - k - p = 1 \end{cases} \quad \begin{array}{l} (0 \leqslant k \leqslant 8,\ 0 \leqslant q \leqslant k) \\ \\ (0 \leqslant k \leqslant 8,\ 0 \leqslant p \leqslant 8-k) \end{array}$$

After solving the equation set above we have one eligible group of (k, p, q)
$$(1, 6, 0).$$

Thus the term $a \cdot b^6 \cdot c$ becomes
$$T(1, 6, 0) = (-1)^6 \cdot {}_8C_1 \cdot {}_{8-1}C_6 \cdot {}_1C_0 \, 3^{8-1-6} \cdot 2^{1-0} \cdot a^{8-1-6} \cdot b^6 \cdot c^{1-0} \cdot d^0.$$
$$= 6 \cdot {}_8C_1 \cdot {}_7C_6 \cdot {}_4C_3 \cdot a \cdot b^6 \cdot c.$$

Its coefficient of the term $a \cdot b^6 \cdot c$ becomes
$$6 \cdot {}_8C_1 \cdot {}_7C_6 \cdot {}_4C_3 = 1{,}344.$$

2.7 Approximate Computation

The binomial theorem is useful in approximate computation of expression of a real number if a certain computation precision (or error) ε is acceptable. For better precision the approximate method discussed below is had better used in the situation where

- the expression can be written to a binomial of an integer and a small decimal, and
- the power n of the binomial is not great.

The procedure of the approximate computation is given as below.

1. Arranging a real number x to a binomial of an integer and a decimal like
$$x^n = (a \pm b)^n$$
(a is an integer and b is a small decimal)
2. Expanding the binomial above using the binomial theorem.
3. Locating a particular term in the expansion, say the $(m+1)^{th}$ term, whose absolute value is less than or equal to the precision (or error) required. The rest after this term can be trimmed.
4. The approximate solution is the sum of the first m terms of the expansion. This method will produce a error less than the absolute value of the $(m+1)^{th}$ term.

Example 2.7.1   *Approximate Computation*

Find approximate value of the following expressions
1) 1.02^4 ($\varepsilon = 0.001$)
2) 0.997^6 ($\varepsilon = 0.001$)
3) 2.995^5 ($\varepsilon = 0.0001$)

Solution:
1) $1.02^4 = (1 + 0.02)^4$
$= 1 + 4 \cdot (0.02) + 6 \cdot (0.02)^2 + 4 \cdot (0.02)^3 + 0.02^4$
$= 1 + 0.08 + 0.0024 + 0.000032 + 0.00000016$

As $\qquad 0.000032 < \varepsilon = 0.001$,
the rest after the 4th term can be trimmed. The sum of the first three terms is the approximate solution.
$$1.02^4 \approx 1 + 0.08 + 0.0024 = 1.0824.$$

2) $0.997^6 = (1 - 0.003)^6 = (1 + (-0.003))^6$
$= 1 + 6 \cdot (-0.003) + 15 \cdot (-0.003)^2 + 20 \cdot (-0.003)^3 + \cdots$
$= 1 - 0.018 + 0.000135 - 0.00000054 + \cdots$

Since $\qquad 0.000135 < \varepsilon = 0.001$,

the rest after the 3rd term can be trimmed. The sum of the first two terms is the approximate solution.

$$0.997^6 \approx 1 - 0.018 = 0.982$$

3) $2.995^5 = (3 - 0.005)^5 = (3 + (-0.005))^5$
$= 3^5 + 5 \cdot (3^4) \cdot (-0.005) + 10 \cdot (3^3) \cdot (0.005^2) + 10 \cdot 3^2 \cdot (-0.005^3) + \cdots$
$= 243 - 2.025 + 0.00675 - 0.00001125 + \cdots$

As $\qquad 0.00001125 < \varepsilon = 0.0001$,

the rest after the 4th term can be trimmed. The sum of the first three terms is the approximate solution.

$$2.995^5 \approx 243 - 2.025 + 0.00675 = 240.98175$$

▶ Simplified Formula of Approximate Computation

From the examples above we found that the sum of the first two or three terms of a binomial expansion can be used as approximate solution for a binomial $(1 \pm x)^n$ when its power n is small. We can use the following two simplified methods in the circumstance where

- $|1 \pm x|$ is close to 1, n is small, and
- the precision ε of the simplified method is acceptable.

Either of the following two simplified methods can be used.

1. Take the sum of the first two terms as the approximate solution of a binomial.

$$(1 \pm x)^n \approx 1 \pm nx \quad (x > 0) \qquad (2.7.1)$$

The error of the approximation computation is close to the absolute value of the third term.

$$\varepsilon = \left| \frac{n(n-1)}{2} \cdot x^2 \right|. \qquad (2.7.2)$$

2. For more precise approximation, take the sum of the first three terms.

$$(1 \pm x)^n \approx 1 \pm nx + \frac{n(n-1)}{2} \cdot x^2. \quad (x > 0) \qquad (2.7.3)$$

2.7 Approximate Computation

The error of the approximation computation closes to the absolute value of the fourth term.

$$\varepsilon = \left| \frac{n(n-1)(n-2)}{6} \cdot x^3 \right|. \tag{2.7.4}$$

Example 2.7.2   *Approximate Computation*

Find approximate solution of the following expressions using simplified approximation formula (2.7.1) and (2.7.3) respectively.
  1) 0.998^6
  2) 1.0005^{10}

Solution:

1) By (2.7.1),
$$0.998^6 = (1 - 0.002)^6 \approx 1 - 6 \cdot 0.002 = 0.988.$$

$$\varepsilon = \left| \frac{6(6-1)}{2} \cdot (-0.002)^2 \right| = 0.00006.$$

By (2.7.3),
$$0.998^6 = (1 - 0.002)^6$$
$$\approx 1 - 6 \cdot 0.002 + \frac{6(6-1)}{2} \cdot (-0.002)^2 = 0.98806.$$

$$\varepsilon = \left| \frac{6(6-1)(6-2)}{6} \cdot (-0.002)^3 \right| = 1.6^{-7}.$$

2) By (2.7.1), $1.0005^{10} = (1 + 0.0005)^{10} \approx 1 + 10 \cdot 0.0005 = 1.005.$

$$\varepsilon = \left| \frac{10(10-1)}{2} \cdot 0.0005^2 \right| = 1.125^{-5}.$$

By (2.7.3), $1.0005^{10} = (1 + 0.0005)^{10}$
$$\approx 1 + 10 \cdot 0.0005 + \frac{10(10-1)}{2} \cdot 0.0005^2$$
$$= 1.005 + 0.00001125 = 1.00501125.$$

$$\varepsilon = \left| \frac{10(10-1)(10-2)}{6} \cdot (0.0005)^3 \right| = 1.5^{-8}.$$

www.ingramcontent.com/pod-product-compliance
Lightning Source LLC
Chambersburg PA
CBHW061219180526
45170CB00003B/1068